AN EARTHED FAITH:
Telling the Story amid the "Anthropocene"
Volume 4

MAKING ROOM FOR THE STORY TO CONTINUE?

Making Room for the Story to Continue?
An Earthed Faith: Telling the Story amid the Anthropocene, volume 4

Originally published by AOSIS Books, an imprint of AOSIS Publishing

Copyright © 2026 Ernst M. Conradie. All rights reserved. Except for brief quotations in critical publications or reviews, no part of this book may be reproduced in any manner without prior written permission from the publisher. Write: Permissions, Wipf and Stock Publishers, 199 W. 8th Ave., Suite 3, Eugene, OR 97401.

Pickwick Publications
An Imprint of Wipf and Stock Publishers
199 W. 8th Ave., Suite 3
Eugene, OR 97401

www.wipfandstock.com

PAPERBACK ISBN: 979-8-3852-7192-4
HARDCOVER ISBN: 979-8-3852-7193-1
EBOOK ISBN: 979-8-3852-7194-8

Cataloguing-in-Publication data:

Names: Conradie, Ernst M. | Vaai, Upon Lumā
Title: Making Room for the Story to Continue?: An earthed faith: telling the story amid the anthropocene, volume 4 / Ernst M. Conradie and Upon Lumā Vaai.
Description: Eugene, OR: Pickwick Publications, 2026. | Making room for the story to continue?: Telling the Story amid the Anthropocene. | Includes bibliographical references and index.
Identifiers: ISBN 979-8-3852-7192-4 (paperback). | ISBN 979-8-3852-7193-1 (hardcover). | ISBN 979-8-3852-7194-8 (ebook).
Subjects: LCSH: Theology. | Human ecology. | Ecotheology. | Trinity. | Storytelling.
Classification: BR115 T10 2026 (print). | BR115 (ebook).

AN EARTHED FAITH:
Telling the Story amid the "Anthropocene"
Volume 4

Making Room for the Story to Continue?

Editors
Ernst M. Conradie
Upolu Lumā Vaai

☙PICKWICK *Publications* • Eugene, Oregon

Theological and Religious Studies editorial board at AOSIS

Chief Commissioning Editor: Scholarly Books
Andries G. van Aarde, MA, DD, PhD, D Litt, South Africa

Board members
Chen Yuehua, Professor of the School of Philosophy, Zhejiang University, Hangzhou, China
Christian Danz, Professor of the Institute for Systematic Theology and Religious Studies, Evangelical Theological Faculty, University of Vienna, Vienna, Austria.
Corneliu C. Simut, Professor of Historical and Systematic Theology, Faculty of Theology, Department of Theology, Music and Social-Humanistic Sciences, Emanuel University of Oradea, Romania; Supervisor of doctorates in Theology, Interdisciplinary Doctoral School, Aurel Vlaicu State University of Arad, Romania; Associate Research Fellow in Dogmatic Theology, Faculty of Theology, Department of Dogmatics and Christian Ethics, University of Pretoria, South Africa.
David D. Grafton, Professor of Islamic Studies and Christian-Muslim Relations, Duncan Black Macdonald Center for the Study of Islam and Christian-Muslim Relations, Hartford International University for Religion and Peace, United States of America.
David Sim, Professor of New Testament Studies, Department Biblical and Early Christian Studies, Australian Catholic University, Australia.
Evangelia G. Dafni, Professor of School of Pastoral and Social Theology, Department of Pastoral and Social Theology and Department of the Bible and Patristic Literature, Faculty of Theology, Aristotle University of Thessaloniki, Greece.
Fundiswa A. Kobo, Professor of Department of Christian Spirituality, Church History and Missiology, University of South Africa, South Africa.
Jean-Claude Loba-Mkole, Professor of Hebrew and Bible Translation, Department of Hebrew, Faculty of Humanities, University of the Free State, South Africa.
Jeanne Hoeft, Dean of Students and Associate Professor of Pastoral Theology and Pastoral Care, Saint Paul School of Theology, United States of America.
Lisanne D'Andrea-Winslow, Professor of Department of Biology and Biochemistry and Department of Biblical and Theological Studies, University of Northwestern-St Paul, Minnesota, United States of America.
Llewellyn Howes, Professor of Department of Greek and Latin Studies, University of Johannesburg, South Africa.
Marcel Sarot, Emeritus Professor of Fundamental Theology, Tilburg School of Catholic Theology: Religion and Practice, Tilburg University, the Netherlands.
Nancy Howell, Professor of Department of Philosophy of Religion, Faculty of Theology and Religion, Saint Paul School of Theology, Kansas City, United States of America.
Piotr Roszak, Professor of Department of Christian Philosophy, Faculty of Theology, Nicolaus Copernicus University, Poland.
Sigríður Guðmarsdóttir, Professor of Department of Theology and Religion, School of Humanities, University of Iceland, Reykjavík, Iceland; Centre for Mission and Global Studies, Faculty of Theology, Diakonia and Leadership Studies, VID Specialized University, Norway.
Wang Xiaochao, Dean of the Institute of Christianity and Cross-Cultural Studies, Zhejiang University, Hangzhou, China.
Warren Carter, LaDonna Kramer Meinders Professor of New Testament, Phillips Theological Seminary, Oklahoma, United States of America.
William R.G. Loader, Emeritus Professor of New Testament, Murdoch University, Western Australia.

Peer-review declaration
The publisher (AOSIS) endorses the South African "National Scholarly Book Publishers Forum Best Practice for Peer-Review of Scholarly Books." The book proposal form was evaluated by our Theological and Religious Studies editorial board. The manuscript underwent an evaluation to compare the level of originality with other published works and was subjected to rigorous two-step peer-review before publication by two technical expert reviewers who did not include the volume editor and were independent of the volume editor, with the identities of the reviewers not revealed to the editor(s) or author(s). The reviewers were independent of the publisher, editor(s), and author(s). The publisher shared feedback on the similarity report and the reviewers' inputs with the manuscript's editor(s) or author(s) to improve the manuscript. Where the reviewers recommended revision and improvements, the editor(s) or author(s) responded adequately to such recommendations. The reviewers commented positively on the scholarly merits of the manuscript and recommended that the book be published.

Research Justification

The traditional theme of God's providence is not often addressed explicitly in the context of Christian ecotheology. Nevertheless, one may find discussions on natural evil, on the theodicy problem, on sustainability, on sustenance (food security), and on multiple threats to life and concerns over the rupture within the Earth System represented by the advent of the "Anthropocene." Each of these raises disturbing questions about any confession of trust in God's loving care. One may also argue that secular discourse on sustainability prompts theological reflection on God's work of sustaining creation despite the impact of anthropogenic ecological destruction (i.e. the impact of "sin"). The title of the volume signals that providence may be understood as curbing the spread of evil to enable God's work of salvation, but the question mark also suggests an uncertainty on whether that may suffice.

This fourth volume in the *An Earthed Faith* series captures the state of the debate in contemporary ecotheology on the theme of providence and extends this debate through a set of diverse constructive contributions from around the world. The ambitious aim of the editorial introduction to this volume is to articulate the core question addressed in this volume, namely, "How could the suffering of God's creatures in the 'Anthropocene' be reconciled with trust in God's loving care?" The literature survey offered demonstrates not only the deep ambiguities embedded in faith in providence but also how this continues to surface in numerous contemporary debates.

The ten contributors (one scholar had to withdraw belatedly) were selected in order to optimize a diversity of positions in terms of geographical context, confessional traditions, and theological schools, also taking considerations of gender, race, age, and language into account.

In "Divine Providence amid Ecological Crises," Aku S. Antombikums explores the theodicy of open theism from within a Nigerian context. In "Common Grace and Sustainability," Ernst M. Conradie unpacks the notion of *conservatio*, noting how this was abused in apartheid South Africa. Nicola Hoggard Creegan offers a view from Aotearoa on the core problem of "Evolution and Providence." Baiju Markose searches for a subaltern planetary ethics in "Divine Providence and Critical Liminalities." In "The Challenge of 'Angry Weather'," Clive Pearson focuses on the existential uncertainties posed by anthropogenic climate change from "down under." Elizabeth Pyne offers an American Catholic perspective on the underlying tensions (or aporia) between "Providence, Suffering, and Creation Faith." In "Mosquitoes, Dengue, and Butterflies," Marisa Strizzi (Argentine) acknowledges that God's providence remains hidden behind a mask. Gloriose Umuziranenge and Eraste Rukera address the suffering imposed by anthropogenic climate change from within a Rwandan context. Finally, Dutch theologian Gijsbert van den Brink raises the question "Does God Care for Cows?," pointing to perils and prospects of faith in God's providence.

The essays included in this volume are all original and develop constructive responses to the same underlying question from within distinct contexts. They adopt a similar methodology, namely a critical review of the available literature in the field of Christian ecotheology, together with constructive argumentation. These are scholarly essays in the sense that they are written by leading scholars, as well as a few emerging scholars in the field. The volume is aimed primarily at experts in the field and has been checked for plagiarism and self-plagiarism. An acknowledgement to the author's PhD thesis was included in "Divine Providence amid Ecological Crises" by Aku S. Antombikums.

The concluding conversation between the contributing authors identifies current paths and emerging horizons on God's providence in order to take the debate forward in the context of contemporary ecotheology. It invites other participants in the field to join the conversation on the basis of this volume.

Ernst M. Conradie, Department of Religion and Theology, University of the Western Cape, Cape Town, South Africa.

Artist Statement

Rev Dr Matagi Jessop Vilitama is a theologian and artist based on his home island, Niue, in Oceania (South Pacific).

The title of the art piece is called "Tafe mai he lagi", which literally means "Poured down from the sky/heavens". It is taken from a Niuean proverb, "Ko e tau monuina tafe mai he lagi" (The abundance flowing from the heavens). The proverb speaks to the times when an individual or the community is blessed with unexpected abundance from the sea or land—yet there is never a time when indigenous Niueans expect any less, given their traditional methods of conservation and preservation. Their deep understanding of the divine providence gives them an assurance of food and pleasure.

The painting reflects the Niuean epistemology on humanity's coexistence with the natural order—that there are patterns and rhythms at work in the natural world. The Moana's (ocean's) tides and currents and abundant sea life—represented by schools of fish on the artwork— is patterned as in a "hiapo" or "tapa" design (traditional bark cloth), transcending the earthly "tutavaha" (horizon). In Oceania cultures, hiapo/tapa is often used to carry gifts on special socio-cultural occasions. The pattern could also be interpreted as the fishermen's net with a bountiful blessing. This hiapo/tapa of fish in this painting symbolises gifts of abundance being rolled down from the heavens, representing God's providence.

Table of Contents

Abbreviations and Acronyms Appearing in the Text and Notes	xv
Notes on Contributors	xvii
An Earthed Faith: Envisaged Volumes in the Series	xxi

"In God We Trust"? A Core Christian Conviction and Its Multiple Challenges amid the "Anthropocene" — 1
Ernst M. Conradie & Upolu Lumā Vaai

An Earthed Faith	1
The Question Posed in this Volume	3
A Set of Core Christian Convictions	4
Challenges Posed by Suffering and Evil	9
Theological Reflection on Providence	13
Providence in Christian Ecotheology	16
Sustenance and Food Security	16
Health and Healing	17
Indigenous Wisdom	17
Ecofeminist Critiques of Patriarchal Care	18
The Universe Story	18
Theologies of Becoming	19
Common Grace	20
Nature Conservation	20
Sustainability	21
Natural "Evil"	22
Structural Violence	22
Progression and Progress in History	23
Marxist and Other Utopias	24
Ecomodernist Optimism	24
Heading Towards a Catastrophe?	25
Discerning God's Finger in Human History	25
A Theology of History	26
Providence and the Critique of Ideology	27
Kenosis or Theosis?	28
Challenges Posed by the "Anthropocene"	28
Bibliography	30

Divine Providence amid Ecological Crises: A Nigerian Perspective — 35
Aku S. Antombikums

Introduction	35
Providence Revisited	37
Classical Theism	38
Wesleyan/Open Theistic View of Divine Control	40
An African Ritualistic View of Creation and Divine Intervention	42
Problems Arising from Current Appropriations of Divine Providence	44
Theocentrism	45
Eschatological Appropriations	47
Domineering Appropriations	49
Divine Providence and the Ecological Crisis: Some Prospects	50
Bibliography	51

Common Grace and Sustainability: Some South African Reformed Perspectives — 55
Ernst M. Conradie

Telling the Story Still Hinges on the Noah Story	55
Neo-Calvinism and its Legacy	59
On Common Grace	60
Common Grace and the Orders of Creation	66
A God of Order in Nature and Society or a God Transforming Society and Nature?	71
Sustainability and Holocene Stability	73
Ecojustice and the Need to Disrupt an Unjust Global Order	75
Van Ruler on God and Chaos	79
Is It God Who Is Stirring the Soup?	82
Theses on God's Work of Conservation	84
Bibliography	86

Evolution and Providence: A View from Aotearoa — 89
Nicola Hoggard Creegan

Introduction	89
Materialism: The Backdrop	92
Scripture	92

Materialist Science	95
Tradition and the Tensions in Theology	97
The "Anthropocene" in Aotearoa / New Zealand	100
Theological Tensions	102
Changes in Biology That Make a Difference	105
Conclusion	107
Bibliography	108

Divine Providence and Critical Liminalities: An Ecotheopoetic Search for Post-Anthropic Subaltern Planetary Ethics — 111
Baiju Markose

Introduction	111
Wood-Pecker	112
Nesting the Ecotheopoetic Imagination	113
Theopoetic Christ: Ornitheological Entanglements	115
Ecotheopoetic Imperative of Liminality in the "Anthropocene"	116
Ecotheopoesis of Subaltern Solastalgic Communities: Divine Providence as the Grounding and Practice of Resistance	118
The Practice of Pārppu: My Grandma's Fugitive Wisdom	119
Arboreal Activism: Kallen Pokkudan's Emplaced Wisdom	121
Towards Subaltern Planetary Ethics	122
Conclusion	123
Bibliography	123

The Challenge of "Angry Weather": Planetary Perspectives on the Providence of God from Life Lived Down Under — 125
Clive Pearson

Situating the Organizing Question	125
Implicating Providence	128
Weather Attribution Science	132
Endgames	134
Intersectionality	136
History Types	139
Revisiting Providence	141
Bibliography	145

On Providence, Suffering, and Creation Faith: Perspectives from "Anthropocene" Aporias   **149**
Elizabeth M. Pyne

Approaching the Question	149
Guiding Insights: Schillebeeckx on "Creation Faith"	151
Amid the "Anthropocene"	158
Everything Has Changed?	158
Nothing Has Changed?	163
Cracks and Questions	164
From the Aporia: Ethics and Providence	169
A Coda on Creation Faith	175
Bibliography	176

Mosquitoes, Dengue, and Butterflies: Providential Masks of the Hidden God   **181**
Marisa Strizzi

Our Share of Suffering	181
The Sum of All Evils	182
The Evils at Home	183
A Theological Fix	185
A Providential Stance	186
God's Hiddenness	187
Truly, Thou Art a God Who Hidest Thyself	187
God Unbound	189
The Suffering of Creation	190
The Nature of Humans	191
Humans and Creation	191
The Hidden Workshop of God	193
The Masks of God	194
Living in the World	194
Providential Masks	196
Here and Now	197
Bibliography	197

God's Providence and Suffering from Climate Change in the African Context   **199**
Gloriose Umuziranenge & Eraste Rukera

Introduction	199
Making a Statement	202

The Practice of Resilience	205
Establishing a Theological Framework	207
Proverbial Sayings	210
Conclusion	210
Bibliography	211

Does God Care for Cows? A Dutch Perspective on the Perils and Prospects of Providence in the "Anthropocene" — 215
Gijsbert van den Brink

Introduction	215
The Perils of Providence in the "Anthropocene"	218
The Particularities of Providence	221
The Prospects of Providence in the "Anthropocene"	230
Bibliography	232

Continuing the Conversation in Christian Ecotheology on God's Providence — 235

Index — 255

Abbreviations and Acronyms Appearing in the Text and Notes

ACT	Association of Consumers and Taxpayers
AGAPE	Alternative Globalization for People and Environment
AIC	African Independent Churches
ASA	American Scientific Affiliation
ATRs	African Traditional Religions
COP	Conference of the Parties
COVID-19	coronavirus disease 2019
CPR	Conseil Protestant du Rwanda
CTI	Center of Theological Inquiry
DNA	deoxyribonucleic acid
EAR	experience of the Anglican Church of Rwanda
EPR	Eglise Presbyterienne au Rwanda
FAO	Food and Agriculture Organization of the United Nations
HIV	human immunodeficiency virus
IPC	Índice de precios al consumidor
IPCC	Intergovernmental Panel on Climate Change
IUGS	International Union of Geological Sciences
KJV	King James Version
NABT	National Association of Biology Teachers
NGO	nongovernmental organizations
NRSVUE	New Revised Standard Version, Updated Edition
PIASS	Protestant Institute of Arts and Social Sciences
RDIS	Rural Development Interdiocesan Service
REET	Ecumenical Network for Theological Education
REMA	Rwanda Environment Management Authority
SACC	South African Council of Churches
SDGs	Sustainable Development Goals

Abbreviations and Acronyms Appearing in the Text and Notes

USA	United States of America
UWC	University of the Western Cape
WCC	World Council of Churches
WSCF	World Student Christian Federation

Notes on Contributors

Aku Stephen Antombikums is a Nigerian Reformed theologian and philosopher interested in the analytic philosophy of religion, analytic theology, and systematic theology. His PhD was in analytic philosophy of religion at the Vrije Universiteit Amsterdam in 2022. He was a postdoctoral researcher at the Vrije Universiteit, Amsterdam. He is currently a research fellow at the Theological University, Utrecht, and a research associate at the University of Pretoria. He has a forthcoming monograph entitled *Divine Control, Human Contingencies and the Problem of Evil* (2025).

ORCID: https://orcid.org/0000-0002-2173-0497
Email: antombikums@gmail.com; aku.antombikums@tuu.nl

Ernst M. Conradie is senior professor in the Department of Religion and Theology at the University of the Western Cape (UWC) in South Africa. He works in the intersection between Christian ecotheology, systematic theology, and ecumenical theology and comes from the Reformed tradition. He is the author of *The Earth in God's Economy: Creation, Salvation and Consummation in Ecological Perspective* (2015), *Redeeming Sin? Social Diagnostics amid Ecological Destruction* (2017), and *Secular Discourse on Sin in the Anthropocene: What's Wrong with the World?* (2020). He was the international convener of the Christian Faith and the Earth project (2007–2014) and co-editor with Hilda Koster of the *T&T Clark Handbook on Christian Theology and Climate Change* (2019). He is the series editor for the project on "An Earthed Faith: Telling the Story amid the 'Anthropocene.'"

ORCID: https://orcid.org/0000-0002-0020-6952
Email: econradie@uwc.ac.za

Nicola Hoggard Creegan is a theologian living in Auckland, Aotearoa (New Zealand). She is the director of New Zealand Christians in Science / Ngā Karaitiana Kimi Matū. She has taught theology in New Zealand and the United States of America (USA). She is the author of *Animal Suffering and the Problem of Evil* (OUP, 2013). She was a participant in the 2012–2013 Center of Theological Inquiry (CTI) year on Human Nature and also participated in the Notre Dame Summer Seminars in Anthropology for Theologians, 2015–2016. She is a research fellow in the Department of Religion and Theology at UWC.

ORCID: https://orcid.org/0000-0001-7698-5270
Email: admin@nzcis.org

Notes on Contributors

Baiju Markose is assistant professor of theology at Trinity Lutheran Seminary in Columbus, Ohio (USA). Previously, he served as a professor of religion and dean of studies at the Dharma Jyothi Vidya Peeth Seminary in Faridabad, India. He is the author of *Notes from the Edges: Theological Intonations* (2021) and *Rhizomatic Reflections: Discourses on Religion and Theology* (2018). He was honored with the Marion McFarland Award by the American Academy of Religion (Midwest Region) in 2017. He is the coordinator of the Global Climate Justice Interfaith Peacemaker Team of the OMNIA Institute for Contextual Leadership, where he works at the intersection of interfaith peace-making and climate justice initiatives. He is registered as a co-researcher at UWC for the project on "An Earthed Faith: Telling the Story amid the 'Anthropocene'."

Email: bjmarkose@gmail.com
ORCID: https://orcid.org/0009-0007-8779-418X

Clive Pearson Clive Pearson is based in Sydney, Australia. He is currently a senior research fellow in the Centre of Religion, Ethics, and Society, Charles Sturt University. His research, teaching, and supervising are in the fields of public, ecological, and cross-cultural theologies, with particular reference to Pasifika, Korean, and local Australian cultures. He has recently stood down as editor of the *International Journal of Public Theology* and has been the editor of a series on Cross Cultural Theologies (Equinox). He is and has been on the editorial board of *Ecotheology*, *Political Theology*, the *Journal of Samoan Theology, and the Korean Presbyterian Journal of Theology*. One of his most recent articles was on whether the "Anthropocene" requires a step-change in theology. He is currently working with Mehmet Ozalp on Christian and Islamic responses to climate change / the "Anthropocene" (Brill). He has supervised theses on theology, climate change and coal, energy, river systems, and low-lying islands. He is a research fellow in the Department of Religion and Theology at UWC.

Email: cpearson@csu.edu.au; cliverpearson@gmail.com
ORCID: https://orcid.org/0000-0003-2557-7562

Elizabeth M. Pyne is assistant professor in the Department of Religious Studies and Philosophy at Mercyhurst University in Erie, Pennsylvania (USA), where she also serves as the director of the William C. Sennett Institute for Mercy and Catholic Studies. Her research addresses issues in systematic, political, and ecological theologies through engagement with critical theory, gender studies, and the environmental humanities. She has published several articles and essays on the work of Edward Schillebeeckx, including a chapter in the *T&T Clark Handbook of Edward Schillebeeckx*

(2019), and has contributed to the collected volumes *Doing Climate Justice* (2022) and *The Environmental Apocalypse: Interdisciplinary Perspectives on the Climate Crisis* (2022). Her first monograph, *Suspended Nature: Theology, Ecology, and Creaturely Politics*, is forthcoming. She is a research fellow in the Department of Religion and Theology at UWC.

ORCID: https://orcid.org/0009-0002-5687-1334
Email: epyne@mercyhurst.edu

Eraste Rukera is the administrative assistant of the dean of the Faculty of Theology at the Protestant University of Rwanda (PUR). He also serves as a minister in the Presbyterian Church in Rwanda. His master's thesis is entitled "An Assessment of the Role of Churches in Mitigating and Adapting Climate Change: A Case Study of the Anglican Diocese of Shyogwe, Huye" (2023). He is registered as a co-researcher at UWC for the project on "An Earthed Faith: Telling the Story amid the 'Anthropocene'."

ORCID: https://orcid.org/0009-0006-7154-2431
Email: rukeraste2@yahoo.fr

Marisa Strizzi Marisa Strizzi is professor and coordinator of the area of theology and interdisciplinary studies in the Ecumenical Network for Theological Education (REET), in Buenos Aires, Argentina. Her areas of interest are theology in postmodernity, theology in relation to ecology, and feminist and queer studies. She is a member of the Mennonite Church community in Buenos Aires. She is the author of *Luther after Derrida: The Deconstructive Drive of Theology* (2022) and co-editor with Dietrich Werner and colleagues of *International Handbook for Creation Care and Eco-Diakonia* (2020). She is registered as a co-researcher at UWC for the project on "An Earthed Faith: Telling the Story amid the 'Anthropocene'."

ORCID: https://orcid.org/0009-0009-5063-2077
Email: mstrizzi@gmx.net

Gloriose Umuziranenge is a senior lecturer and the director of quality assurance at the Protestant University of Rwanda (PUR). She holds a PhD degree in human geography from the University of Bamberg in Germany. She contributed to the establishment of the Department of Natural Resources and Environmental Management at PIASS, and she was the first ever head of this department. Her research interests and publications are related to community participation in natural resources management and governance, environmental justice, and ecotourism in the Rwandan context. She is a research fellow in the Department of Religion and Theology at UWC.

ORCID: https://orcid.org/0000-0001-5430-8121
Email: u.gloriose@gmail.com

Notes on Contributors

Upolu Lumā Vaai (PhD) is professor of theology and ethics and principal of the Pacific Theological College in Suva, Fiji. He works in the intersection between Pacific contextual decolonial theology and Trinitarian systematic theology within the context of an Indigenous philosophy of relationality and how this could be used as a decolonial tool to liberate communities from both past and current dominant imperial policies and models of development. His recent co-edited publications include *Restoring Identities: The Contextual Story of Christianity in Oceania* (2023), *reStorying the Pasifika Household* (2023), and *Methodist Revolutions* (2022). He is registered as a co-researcher at UWC for the project on "An Earthed Faith: Telling the Story amid the 'Anthropocene'."

ORCID: https://orcid.org/0000-0003-1243-0852
Email: ulvaai@ptc.ac.fj

Gijsbert van den Brink is professor at the Faculty of Religion and Theology of the Vrije Universiteit Amsterdam in the Netherlands. He is a systematic theologian from the Reformed tradition, and his research is focused on the intersection of Christian theology and the sciences. In particular, he participates in debates on theology and evolution, the future of academic theology, and (more recently) ecotheological themes. He is the author of *Philosophy of Science for Theologians* (2009), *Christian Dogmatics: An Introduction* (2017; co-authored with Cornelis van der Kooi), *Reformed Theology and Evolutionary Theory* (2020), *Dawn: A Proton's Tale of All That Came to Be* (2021; co-authored with Corien Oranje en Cees Dekker), and *Test All Things: The Bible, Faith and Science* (2023). He is co-editor of *Progress in Theology: Does the Queen of the Sciences Advance?* (2024). He is an extraordinary researcher at the Unit for Reformational Theology and the Development of the South African Society, Faculty of Theology, North-West University, South Africa.

ORCID: https://orcid.org/0000-0002-6950-734X
Email: g.vanden.brink@vu.nl

An Earthed Faith: Envisaged Volumes in the Series

The following twelve volumes are envisaged in the series entitled "An Earthed Faith: Telling the Story amid the 'Anthropocene'":

1. **Taking a Deep Breath for the Story to Begin ... An Earthed Faith 1 (Prolegomena)**

This volume will address the following question: How does the story of who the Triune God is and what this God does relate to the story of life on earth? Is the Christian story part of the earth's story or is the earth's story part of God's story, from creation to consummation? This raises many issues on the relatedness of religion and theology, the place of theology in multidisciplinary collaboration, the notion of revelation, the possibility of knowledge of God, hermeneutics, the difference between natural theology and a theology of nature, and so on. The word "breath" in the title suggests the Spirit of God as source of inspiration for the story, already present in any further deliberations. It hints at an air of anticipation, indicated by the three dots in the title.

2. **How Would We Know What God Is Up To? An Earthed Faith 2 (Method)**

This volume will address the following question: Given what we know about the "Anthropocene," how does one even begin to answer the question of what is this God up to? And how would we know how to respond to that? These are questions of theological method, including the sources and interlocutors of Christian theology, its aims and starting points, social theories shaping it, and presuppositions grounding it. Addressing this question is the classic task of doing contextual theology, namely to describe and analyze the particular context that is addressed and to consider how this may best be addressed theologically. This question highlights the need for prophetic theology to discern the "signs of the times," to recognize a "moment of truth" (*kairos*), and to discern counter-movements of the Spirit. Such methodological questions are necessary in order to tell the story of who God is and what God does amid the "Anthropocene." In terms of narrative/rhetorical theory, a focus on method requires attention to the plot upon which the narrative hinges; the sense of crisis that will draw together the characters; the exigencies that invite passion, reflection, and persuasion. Theological method is inherently a theological question, about sin and salvation, creation and redemption, God and God's world—and shapes where the story may lead and how it may be told.

3. The Place of Story and the Story of Place? An Earthed Faith 3 (Creation)

This volume will address the following question: What difference does it make to the story of cosmic, planetary, human, and cultural evolution to re-describe this as the creative work of God's love? Inversely, what difference does it make to the story of God's love to describe it in evolutionary terms? Addressing this question will require theological reflection on creation and cosmic, biological, hominid, and human evolution (the story of place). Such reflection on the beginning is, of course, not situated "in the beginning" but entails a narrative reconstruction of the story where current interests, positions of power, and fears are necessarily at stake (the place where the story is being told). This is a contested space, indeed a "site of struggle," often dominated by issues of race rather than by grace. How, then, is this story to be told given a sense of place? It will not be possible to avoid questions around suffering, sin, evil, and the tragic (the theme of the next volume), but the focus will be on why on earth a loving God would deem this story to be "very good"—despite the prevalence of suffering, injustice, and oppression.

4. Making Room for the Story to Continue? An Earthed Faith 4 (Providence)

This volume will address the following question: How could the suffering of God's creatures in the 'Anthropocene' be reconciled with trust in God's loving care? Addressing this question will require theological reflection on the classic themes related to the doctrine of providence, including *creatio continua*, *conservatio*, *gubernatio*, and *concursus*. For some, God's providence (common grace) is a necessary requirement to allow (to make room for) the history of salvation to proceed. For others, the suffering embedded in God's "good" creation requires responses to the theodicy problem: why would a loving God allow creatures to suffer so much? What is the relationship between so-called natural evil and social evil? Is the underlying problem human sin, or is it the inadequacies, the tragic dimension, indeed, the violence embedded in God's world? Again, this last question is hinted at in the question mark after the title.

5. The Saving Grace of the Story? An Earthed Faith 5 (Soteriology)

This volume will address the following question: How is the Christian message of salvation to be interpreted given current ecological destruction and apocalyptic fears associated with the "Anthropocene"? Is this message plausible given the failure of Christianity to address so many other urgent problems over 20 centuries? This will require theological reflection on Christological symbols such as atonement and Pneumatological symbols such as liberation, healing, reconciliation, regeneration, moral guidance, justification, and sanctification—insofar as these may be pertinent in the age of the "Anthropocene." The title is ambiguous and ironic to indicate that the story is highly contested but is, at best, to be understood as good news for the whole earth.

6. The Keepers of the Story? An Earthed Faith 6 (Ecclesiology)

This volume will address the following question: What is the place and significance of the church in God's "household," now situated in the destabilizing context of the "Anthropocene"? Addressing this question will require theological reflection on the formation, up-building, and very nature of the church, on its many ministries and missions. Presumably, the question is no longer whether there is salvation outside of the church, but indeed whether there is salvation to be found within the church. Can it still be said that the church is God's main (even only) instrument (sign, sacrament, icon) to bring salvation, given the challenges posed by the "Anthropocene"? Or is the task of the church the monastic one of "keeping" the story, that is, to maintain the inner secret to the mystery of history, amid dark clouds looming and despite few outsiders taking any notice? Does this not sound as if it is the church that needs to come to God's rescue, or is the inverse true?

7. Where the Story Ends and Its Ends ... An Earthed Faith 7 (Eschatology)

This volume will address the following question: How should the content and significance of Christian hope be understood in the context of the "Anthropocene"? Addressing this question will require theological reflection on the eschatological symbols of the final judgement as a sign of hope, on the resurrection of the dead, on the coming reign of God, and on eternal life. It will also have to assess whether such hope is to be understood as the restoration (neo-Calvinism), elevation (Roman Catholicism), replacement (Anabaptism), recycling (liberalism/secularism), or deification/theosis (Eastern Orthodoxy) of this world. Does the meaning of the story lie in its end, or in the journey/pilgrimage towards that end? Any answer to such questions will remain provisional, hinted at in the three dots in the title.

8. Being Blessed as the Inner Logic of the Story? An Earthed Faith 8 (Election)

This volume will address the following question: Can the notion of being God's chosen people/instrument be retained in a religiously plural world under the threat of the "Anthropocene"? Addressing this question will require theological reflection on the themes of divine election and vocation. Can "being blessed" by God be understood as the inner logic of the story? Is such blessing not often experienced as a curse? What about divine reprobation, punishment, and justice for the victims and perpetrators of history? How is a theology of religions to be understood in a context characterised by common threats and the need for tolerance and compassion across religious divides? How can Christians move beyond the options of exclusivism and relativism in the context of the "Anthropocene"? What does it mean to be blessed, for the whole of creation to receive God's blessing?

9. The Spirit of the Story? An Earthed Faith 9 (Pneumatology)

This volume will address questions around the identity and character of God's Spirit. It will require theological reflection on how the very notion of spirit should be understood in relation to person, matter, ideas, force, energy, and related concepts. What does it mean that this Spirit is "holy" and makes things "holy"? Is this Spirit able to overcome what is "demonic" in the "Anthropocene"? Is it money or love that makes the world go round? Or is this Spirit the spirit that makes matter move, even if this movement is not all that obvious and requires discernment?

10. The Letter of the Story? An Earthed Faith 10 (Christology)

This volume will address questions around the identity and character of Jesus of Nazareth, proclaimed to be the Christ, anointed by God's Spirit, the One who would inaugurate God's coming reign. It will require theological reflection on the significance of all six Christological symbols, namely (deep) Incarnation, Cross, Resurrection, Ascension, Session, and Parousia as these may relate to the coming of the "Anthropocene." If the cross is a concrete symbol of the history of imperialism and oppression, can the (bodily?) resurrection still function as an equally concrete symbol of hope in the "Anthropocene"? How is the interplay between the letter and the spirit of the story to be understood given long-standing ecumenical divides on the *filioque* controversy—which still divides the East and the West, the North and the South—over whether the Spirit works (only/primarily) on the basis of the Letter (as most so-called mainline churches assume)? Or should the relative independence of God's Spirit be emphasized (as many others emphasize)?

11. In Communion with the Storyteller(s)? An Earthed Faith 11 (Trinity)

This volume will address questions around the doctrine of the Trinity as the inner secret / apophatic Mystery / doxological culmination of the Christian faith. It will offer theological reflection on how the economic Trinity and the immanent Trinity are related by exploring God's identity and character. The question is which of God's characteristics need to be foregrounded in the age of the "Anthropocene." In particular, how is God's mercy related to God's justice given the interactions between God as Father, Son, and Spirit? Can these (patriarchal) symbols be maintained in the "Anthropocene"? Should one favor the social analogy (emphasizing communion) or the psychological analogy (perhaps allowing for a more generic notion of God) for understanding the Trinity? What difference does faith in such a God make (if any) in the age of the "Anthropocene"? Moreover, who is telling the story? Are we (Christians?) the ones responsible to tell the story or are we characters in a story ultimately told by Godself? Given these reflections, what does it mean to believe in "God" (a God, any God) in the world in which we now live? Note that this (philosophical) question is not addressed upfront but penultimately. For Christians, the question remains whether this Triune God can be regarded as the ultimate mystery of the world.

12. What, Then, Is the Moral of the Story? An Earthed Faith 12 (Ethics)

This volume will address questions around the relationship between Christian doctrine, Christian ethics, Christian spirituality, and Christian praxis—between the ultimate and the penultimate, between the indicative of God's grace and the imperative of ecological gratitude. Such relatedness has been there implicitly in all the other volumes but needs to be made explicit here. In dealing with climate change (for example), there is a need to find common moral ground with those standing in other religious traditions and with organizations in civil society. This has implications for all the relevant ethical categories—such as moral vision, virtues, duties, rights, responsibilities, values, middle axioms, action steps, and so on. For Christians, the question will be whether (and if so, how) such common moral ground is deeply rooted in the story of who God is and what God has done, is doing, and will be doing.

"In God We Trust"? A Core Christian Conviction and Its Multiple Challenges amid the "Anthropocene"

Ernst M. Conradie[1] & Upolu Lumā Vaai[2]

■ An Earthed Faith

This is the fourth volume in the series entitled *An Earthed Faith: Telling the Story amid the "Anthropocene."* The aim of this series is to offer collaborative, constructive contributions to understanding the content and significance of central themes of the Christian faith from perspectives in Christian ecotheology, given the challenges associated with the so-called "Anthropocene."

---

1. Ernst M. Conradie is senior professor in the Department of Religion and Theology at the University of the Western Cape in South Africa.

2. Upolu Lumā Vaai is professor of theology and ethics and Principal of the Pacific Theological College in Suva, Fiji.

> **How to cite:** Conradie, EM & Vaai, UL 2024, '"In God We Trust"? A Core Christian Conviction and Its Multiple Challenges amid the "Anthropocene"', in EM Conradie & UL Vaai (eds.), *Making Room for the Story to Continue?*, An Earthed Faith: Telling the Story amid the "Anthropocene", vol. 4, AOSIS Books, Cape Town, pp. 1-33. https://doi.org/10.4102/aosis.2024.BK415.01

The assumption is that the Christian faith has a narrative shape and structure. It tells a story of who God is and what God has done, is doing, and might be doing within our world. The question "Who is God?" is answered by telling a story (e.g., Deut 26:5–9). Clearly there are different versions of this story, multiple perspectives on the story, and conflicting images and understandings of who God is, what God's identity and character may be, how we could possibly know that, and indeed what being divine may mean. What God may be doing is likewise contested. There are also many strands of narrative theology. But this should not be allowed to confuse the issue: it matters which God/god one is talking about. Somehow these stories and perspectives on Christian faith need to touch each other, unless there are different gods altogether, or unless God's actions are in conflict with each other, for example by suggesting that redemption is necessary to overcome the inadequacies of God's work as Creator. The definite article in "telling the story" is clearly contested too. Nevertheless, to use the indefinite article ("a Christian story") or the plural ("Christian stories"), without linking the various aspects of God's work to each other, runs the risk of allowing the various aspects of God's work to become disentangled, God's Triune identity to disintegrate into an amorphous polytheism, and Christian witnesses to be relativized in advance.

Throughout the ambiguous history of Christianity, it was far from easy to hold these together, and attempts at doing so readily became hegemonic. One therefore finds, *de facto*, not one Christian tradition but many, not one faith but conflicting Christian faiths, not one history but contested histories, not a united but a fragmented Christianity.

Moreover, the contributions to this series are situated "amid the Anthropocene" with the recognition that a "business as usual" way of doing theology is no longer appropriate. There is no need here to offer introductory comments on the "Anthropocene," as the multidisciplinary literature in this regard is rapidly expanding.[3] This will come into play in many of the essays included in this volume. Although the term "Anthropocene" is itself heavily contested,[4] it is at least clear that we live in a time where the balance

---

3. See the attempt to make theological sense of this debate by Ernst Conradie in "Some Theological Reflections on Multi-disciplinary Discourse on the 'Anthropocene'."

4. An increasing number of scholars in various disciplines raise questions about the use of this term because it obscures the particularity inherent in anthropogenic causation and in responsibility. By suggesting "humans" as the causal factor, the term may cloud over the reality that some humans are far more implicated than others in historical and contemporary greenhouse gas emissions, and that the lines of causation are highly racialized and class dependent. For the *An Earthed Faith* series, it was decided to always use the term "Anthropocene" in quotation marks to indicate the contestations over naming it as such. Doing ecotheology "amid" the "Anthropocene" is then not only a reference to a rupture in the Earth System but also to the dominant ways of interpreting it. Following the decision by the International Union of Geological Sciences on March 4, 2024 to *reject* the proposal from the Anthropocene Working

between Earth subsystems has become disturbed (or "ruptured"),[5] where there is a shift away from the relative stability that characterized the Holocene—and where (Western) Christianity stands accused by many as one of the deepest causes of the underlying problem. How, then, should this story of who God is and what God is doing be told?

Volume 1 in the *An Earthed Faith* series offered a survey of the many strands of narrative theology[6] and then posed the question of how the Christian story of who God is and what God has done relates to the story of the universe. Volume 2 explored the emergence of ecotheology as a scholarly discourse, its global spread and fragmentation,[7] in order to raise questions around an appropriate method or methods of doing ecotheology. The third volume explored shifting agendas in creation theology within the context of Christian ecotheology in order to fathom a sense of place from which the story of God's work is told—and the contested story of each place.[8] This fourth volume will explore God's providence as a second dimension of God's work, alongside that of creation. In whatever way this may be construed, God's work of providence forms an integral part of the story of who the Triune God is and what this God has done, is doing and may be doing amid the "Anthropocene."

■ The Question Posed in this Volume

In each volume in the series, a single question is raised to which contributors are asked to respond from within their particular context and confessional tradition. Given this pattern, the fourth volume addresses the following core question:

> How could the suffering of God's creatures in the "Anthropocene" be reconciled with trust in God's loving care?

Addressing this question will require theological reflection on the classic themes related to the doctrine of providence, including *creatio continua*, *conservatio*, *gubernatio*, and *concursus*. For some, God's providence

---

Group to establish the Anthropocene as a formal epoch in the earth's geological timetable, despite ongoing criticism, it would be inappropriate to refer to the "Anthropocene Epoch." Nevertheless, the term will still be widely used in scientific studies, in the humanities, and in popular debates to describe human-induced change to the Earth System.

5. See Hamilton, "Anthropocene as Rupture."

6. See the introductory essay by Ernst Conradie with Lai Pan-Chiu, entitled "On Setting the Scene for the Story to Begin."

7. See the introductory essay by Ernst Conradie and Cynthia Moe-Lobeda, entitled "Telling the Story *en route*: On this Road (*hodos*) and its Logic (*logos*)."

8. See the introductory essay by Ernst Conradie and Willie James Jennings, entitled "The Place of Creation in Christian Ecotheology—Some Shifts in the Story."

(common grace) is a necessary requirement to allow (to make room for) the history of salvation to proceed. For others, the suffering embedded in God's "good" creation requires responses to the theodicy problem: why would a loving God allow creatures to suffer so much? What is the relationship between so-called natural evil and social evil? Is the underlying problem human sin, or is it the inadequacies, the tragic dimension, indeed the violence embedded in God's world? This last question is hinted at by the question mark after the title.

The ten contributors to this volume were selected in order to optimize a diversity of positions in terms of geographical context, confessional traditions, and theological schools, also taking into account considerations of gender, race, age, and language (to address distortions that may result from using English as a medium of communication). The aim of the series is ambitious, namely to discern current paths in the state of the debate in Christian ecotheology, to identify emerging horizons in the field, to take the debate forward through a set of constructive contributions to current discourse, and, on this basis, to open up the field for further contributions from around the world. While this introductory essay needs to paint in broad strokes in order to offer a proverbial bird's eye view of such current paths, the ten constructive contributions that follow will each speak from a particular place in order to take the debate forward.

In order to stimulate such further discussions, this introductory essay offers some reflections on how the theme of providence has been approached in the specific context of Christian ecotheology. It will become clear that the core Christian conviction expressed in the notion of providence, namely that God cares for us, soon becomes contested, confused, and conflated in theological discourse. This can only obscure and undermine the comforting assurance that God is like a caring parent. It is therefore important not to lose sight of the roots of this conviction in Christian piety.

∎ A Set of Core Christian Convictions

All the contributors to this volume observe that there is a tendency in contemporary Christian theology to avoid the notion of providence. It is often not treated as an aspect of God's work in its own right. One reason may be that it is replaced by a discussion of the theodicy problem (see below), which is itself often criticized as a fruitless endeavor. Another reason is that classic notions of providence seem to be trapped in categories derived from Greek philosophy that constrain further reflection and can hardly be related to Indigenous wisdom elsewhere in the world. The reason may well be that providence (Latin: *providere*) is understood as foresight and is then linked to God's omniscience. If so, this leads to an emphasis on a rational world order, raises questions around divine determinism, and underplays human and other forms of agency.

Nevertheless, traditionally, the doctrine of providence refers to reflection on a core Christian conviction, namely that God cares for and provides for our human needs. This is assumed in the two connotations embedded in the word providence, namely foresight (again *providere*) and provision.[9] God cares not only for the righteous but for all people, even the wicked. Such care is also evident in the food provided for animals, for the birds in the air (Matt 6:26), for cedar trees, wildflowers, and all that lives (Ps 104). Indeed, even though five sparrows are sold for two pennies, not one of them is forgotten in God's sight (Luke 12:6).

In the context of ecological destruction, now epitomized by the so-called "Anthropocene," such care is readily extended to the whole world, indeed to the whole of God's beloved creation (as per John 3:16a). A trust in God's care is exemplified by Psalm 23 ("The Lord is my shepherd; I shall not want") and Psalm 121 ("My help comes from the Lord, who made heaven and earth"), both of which recognize stark surrounding challenges.

This conviction that God cares shapes the daily life of the Christian laity. It is expressed in petitionary prayers for food; health and well-being; fertility; shelter from cold, heat, and wind; for rains on time and a good harvest; safety on a journey; or for protection from multiple dangers (e.g., the elements, tides and floods, darkness, predators, attackers, enemies). The trust in God's care thus expresses a recognition in prayer of what lies ultimately beyond the locus of control of an individual human agent, family or community. Accordingly, Christians may find consolation in Romans 8:28: "We know that all things work together for good for those who love God, who are called according to his purpose" (New Revised Standard Version, Updated Edition [NRSVUE]). In short: if God is for us, who can be against us (Rom 8:31)? One may therefore say that providence is best understood in the context of a life of prayer and in doxology; it is otherwise easily misunderstood and then just does not make sense.[10]

It is perhaps especially in Indigenous communities where such a trust in God's care thrives because of a sense of vulnerability to destructive forces beyond their control. Such Indigenous communities are found around the world, but prime examples include small island states that are vulnerable to the effect of climate change. By contrast, in affluent communities, an expression of trust in God's care (e.g., for the food on the table) can easily revert to a legitimation of an upper-middle class lifestyle as "proof" of

---

9. See Fergusson, *The Providence of God*, 298.

10. See Fergusson, *The Providence of God*, 322–30.

God's tender loving care. The danger here is one of deducing from prosperity that this is a sign of God's "many blessings." Adversity may then be identified with God's curse[11] and prosperity with salvation. The very same conviction may therefore serve the purpose of self-legitimation in times of privilege and power. This would underestimate the biblical critique of local fertility and prosperity cults (the Ba'alim) in the name of Yahweh, the God of Israel. The prosperity gospel thrives not only among the consumer class but also among the poor and lower-middle class who hope that their Christian adherence will yield some much-needed prosperity.[12]

Alongside such a need for sustenance, trust in God's care also lives in the hearts of believers who desire guidance in the education of their children, important life choices (marriage, career, moving house, migration) and wisdom in coping with the many demands of life. The same would apply to institutions and their leaders in need of wisdom for policy-making and decision-making, given that the full complexity of the present situation cannot be grasped, while what the future may hold always contains an element of uncertainty, at times radical uncertainty. Such guidance cannot be reduced to divine commands but may be expressed in convictions on God's presence in the life of individual believers, families, congregations, other institutions, nations, and even the world as such. Accordingly, God's word is a lamp for one's path in life (Ps 119:105).

Such guidance may be extended to the conviction that, ultimately, it is not our decisions that determine our lives, but that we may leave the outcome in God's hands. This is radicalized in the trust that God is able to turn experiences of suffering and evil (which God did not cause) towards the good, thus using them as opportunities for grace. Prime biblical examples of such a *Gesetz der Umlenkung* include Joseph being sold into slavery by his own brothers, Emperor Cyrus of Persia being employed as God's servant, and the execution of Jesus Christ on a Roman cross that God inverts for the sake of the salvation of the whole world. From this conviction emerges the trust that God somehow governs our lives, families, communities, congregations, institutions, even nations. Accordingly, God has a purpose for us and every other creature—and will bring that to fulfillment despite many experiences that point to the contrary. One can therefore detect God's hand in the myriad of details in one's personal life.

This conviction may be extended towards (contemporary) history as such. Accordingly, God is steering history towards God's ultimate goal for the

---

11. See Berkouwer, *The Providence of God*, 179.

12. There is a significant corpus of literature on the prosperity gospel from around the world. For a University of the Western Cape–based discussion with students from such contexts, see Conradie, *Christianity and a Critique of Consumerism*.

world, despite multiple deviations, distortions, and disruptions. One may discern a sense of direction in history towards the coming reign of God, whether at a personal, ecclesial, national, or global level. Ultimately, history is in God's hands. If so, the theological question is what God is up to at this moment in history, again in my life and in this community, in widening circles.[13] In ecotheology, a cosmic extension of any such notion of providence is needed: one has to affirm that God's faithfulness is not only to God's covenant (with Israel) but to the whole of creation, despite the fall of humanity.

On this basis, Christians are encouraged to seek and find traces of the victory of the resurrected Christ in contemporary events. As Hendrikus Berkhof puts it, "Faith does not depend on vision, but it does lead to vision. For faith, among other things, means the certainty that wherever Christ is glorified in this world and reigns over the world, his resurrection power is active in history and as such takes on form, i.e., it becomes visible."[14]

One may conclude that trust in God's providence yields the virtues of humility and gratitude, patience in adversity, and confidence amid worries about things that one cannot control. It need not foster Stoic resignation or moral paralysis but elicits protest and resistance on the borderline between what is and what is not within one's locus of control. If everything is in the powerful yet loving hands of God, then humans can face any external calamity and internal weakness with serenity, courage, and confidence.[15] If, by contrast, we live in a world governed by either fate or pure randomness, we are driven to unrelieved despair.[16] In response to the question of how believers are helped by faith in God's providence, the Heidelberg Catechism responds:

> We can be patient when things go against us, thankful when things go well, and for the future we can have good confidence in our faithful God and Father that nothing in creation will separate us from his love. For all creatures are so completely in God's hand that without his will they can neither move nor be moved.[17]

---

13. See Charles Wood's proposal that the doctrine of providence responds to the question what is going on from a theological perspective. See Wood, *The Question of Providence*, 12.

14. See Berkhof, *Christ the Meaning of History*, 133.

15. See Gilkey, *Reaping the Whirlwind*, 177, drawing on John Calvin.

16. See Gilkey, *Reaping the Whirlwind*, 178-79: "In a world governed by the former [fate and chance] all happens blindly, with no meaning at all; here there is promise of neither liberation nor salvation. Since we are in such a world subjected to utterly senseless forces, all the miseries that these forces bestow on us, i.e., of failure, disease, bereavement, death, have no meaning vis-à-vis our own hopes and thus insofar as they are victorious—and they frequently represent an unrelieved despair. Here, therefore, there is the possibility of neither serenity nor consolation and no ground for courage in facing life's inevitable trials. At best we can depend only on our own transient and fragile powers to elude the forces that menace us—and thus do we face the inner risk, if we succeed, of callous pride, and, if we fail, of utter despair."

17. Question and Answer 28. For this translation of the Heidelberg Catechism, see https://www.crcna.org/welcome/beliefs/confessions/heidelberg-catechism#toc-god-the-father [last accessed October 9, 2024].

Providence thus requires contentment with what one cannot change in life. In the famous, widely cited prayer for serenity attributed to Reinhold Niebuhr, knowing the difference between what lies within one's locus of control and what doesn't is what matters: "God, grant me the serenity to accept the things I cannot change, the courage to change the things I can, and the wisdom to know the difference."

Note that in a Christian context, providence is not to be equated with destiny or fatalism, nor with a sense of karma that suggests that one inevitably reaps what one sows, that one gets what one deserves. That can only encourage passivity. A belief in providence is in fact a protest statement against vulnerability to the whims of luck or the fatalities of fate. The inverse of determinism, namely a belief in luck or chance (and the many superstitions associated with that), also cannot be attributed to God's providence. Likewise, providence cannot be reduced to what has been allotted to us in life in terms of length and strength, intelligence, good looks, musical ability, or physical agility. This is how "Providence" (with a capital) is sometimes understood in a secular context and thus becomes a nickname for God (also used by Adolf Hitler!). Another distortion of God's care (due to Greek influence) is that providence (the Latin *providere* means seeing ahead in the sense of foresight) may be equated with divine foreknowledge or with divine predetermination, so that providence and fate become merged with each other.[18]

In response to secular notions of "Providence" that tend to become deist, contemporary Christian treatments of God's providence seek to hold together God's work of creation, providence, and salvation more closely. There is a recognition of the need for a more Trinitarian understanding of God's care based on the recognition that the Father who cares is none other than the Father of Jesus the Messiah—raising the question why the Father did not intervene when the Son was crucified. Moreover, the Spirit is a dynamic power of life (the Hebrew *ruach*), a motherly comforter, who proceeds from the Father and upon whom all of life is dependent, so that God's care precedes, indwells, and enables human care. God's care is then best associated with the Spirit as Advocate and Comforter.[19] This allows for a less deterministic (Stoic) and a more interactive notion of providence. Put differently, providence is less about God's lordship over all things (God's foreknowledge, sovereignty, and thus power),[20] being in control of all that happens, and more about God's vulnerable love.

---

18. See Wood, *The Question of Providence*, 24.

19. See, e.g. Fergusson, *Creation*, 50-61, also his *The Providence of God*.

20. See Gorringe, *God's Theatre*, 5.

■ Challenges Posed by Suffering and Evil

Such a trust in God's care emerges precisely from experiences of its apparent absence. When one has sufficient food, health, rain, harvests, shelter, and safety, one may well take such things for granted. It is from concerns, uncertainties, and anxieties, when these things are threatened, that a trust in God's care becomes meaningful. It is during times of political uncertainty and upheaval that God's care provides assurance.[21] Counter-experiences of suffering do not falsify but may actually elicit such trust. Likewise, providence is not contradicted by the impact of human sin but is precisely made necessary as a result of sin.[22] In fact, an affirmation of God's providence emerges and is tested by being confronted with "principalities and powers"; with distortive ideologies; with economic, political, and legal corruption; and with forms of tyranny and state capture. As a result, God's providence typically remains hidden in the contrast between what is and what should be (God's purposes). The "Hallelujah!" that reverberates in singing God's glory in any doxology all too easily turns into a triumphalist theology of glory to justify power and privilege, unless it is expressed as a protest against oppressive powers.[23] The triumphant chorus, "For thine is the kingdom, and the power, and the glory, for ever. Amen." in the Lord's Prayer follows upon "And lead us not into temptation, but deliver us from evil" (Matt 6:13, King James Version [KJV]).

However, when suffering becomes too much to bear, believers are left to wonder whether God really cares. Does God actually care? Can God care? Does it even make sense to say that God cares? Does this conviction not die the death of a thousand qualifications, by every experience of unjustifiable suffering? Moreover, is trusting in God's care not dangerous and open to abuse and ideological distortion? As David Fergusson observes, it was an intuitive assumption of nineteenth-century imperial rhetoric that empires were providentially ordained to transmit the benefits of Western civilization to other parts of the world.[24] "In God We Trust" is the official motto of the United States of America, adapted from Psalm 91:2 and adopted by the U.S. Congress in 1956. It appears on all American currency so that trust in God's care is easily conflated with trust in the undoubted power of the U.S. dollar.

---

21. David Fergusson, following insights from Heiko Oberman, refers to John Calvin's status as a refugee in Geneva to account for the profound role (with lurking dangers) that providence plays in a Reformed spirituality. See *The Providence of God*, 77–81.

22. See Gilkey, *Reaping the Whirlwind*, 161.

23. See Berkouwer, *The Providence of God*, 267.

24. See Fergusson, *The Providence of God*, 3.

Given the destructive ecological impact of industrialized capitalism, now evident in the "Anthropocene" (or, as some insist, the "Capitalocene"),[25] this may well constitute a heretical betrayal of a God who cares for those escaping slavery, oppressed peoples, and those marginalized by society, as well as by religious authorities. Those who have become the victims of modernization, globalization, and American hegemony (and that includes multiple other forms of life) can hardly be asked to affirm "In God We Trust" as their own motto. Notions of God's sovereignty, where affirmed by the powerful, can only be regarded as a legitimation of power by others. Functionally replacing a trust in God's care with trust in the American dollar is nothing but a heretical distortion of a core Christian conviction. What some regard as a Day of Thanksgiving for God's providential care, others dismiss as colonial conquest. The problem remains that Christianity is so closely related to colonialism and capitalism (three C's) that no notion of God's governance in history can escape from this legacy.

Moreover, does a sense of the tragic not pervade our lives? Is God still somehow in control of where things are going towards, or are we (some of us humans) now in control? Or is our world determined by something else, astral fate or luck, fortune or chance, perhaps? Is it not easier and more consoling to attribute suffering to accidents, misfortunes, negligence, or errors that could be rectified by a sacrifice, a gift, or even a bribe? More pertinently: does God's professed care make any difference to human and animal suffering? Or is this nothing more than a decorative way of talking about finite human capabilities, perhaps pious and humble, perhaps self-deceptive, even operating as the opium *of* the people (Marx) or *for* the people (Lenin)? Such doubts about God's care are often expressed in the biblical witnesses themselves, notably in the Psalms, in the Book of Job, in the Lamentations, and also by Jesus the Christ on the cross: "My God, why have you abandoned me?" (Matt 27:46).

---

25. See, for example, Joerg Rieger's *Theology in the Capitalocene*, following Jason Moore's *Anthropocene or Capitalocene*. Rieger insists that this is an age in which a small, privileged group's economic interests and emphasis on maximising profits rule both people and planet (2–3). While the quip that it is easier to imagine the end of the world than the end of capitalism rings true, the alternatives of socialism or communism (as responses to industrialized capitalism) focus on the distribution of wealth but still assume the need for sustained economic growth and maintain a similar carbon footprint—and in that sense remain part of the problem. The geological impact of humans is intertwined with the history of capital but will also be evident long after capitalism has gone or morphed into something else. The history of capitalism is therefore not by itself sufficient to understand the human predicament (see Chakrabarty, *The Climate of History in a Planetary Age*, 4, 35). It is unlikely that something as complex as the "Anthropocene" can be attributed to a single cause, even if its impact is as pervasive as that of capitalism. The question is also how to understand the anti-colonial quest for modernization in so-called developing countries and why those who do not form part of the small, privileged group (the "global 99 percent") are lured by the consumerist dream of prosperity. For Rieger, the answer lies in the capitalist "treadmill of production" based on endless accumulation and requiring an advertising industry to maintain. Sin is understood as capitalist exploitation of people and earth by the few, rather than the consumerist greed of the many (188).

Typically, believers would distinguish between natural causes of suffering (earthquakes, volcanoes, bad weather, accidents, droughts, famine, sickness, pests, aging) and social causes of such suffering in the form of injustices, oppression, even genocide—where suffering is the result of what some humans and their institutions do to others, including other forms of life. Here, a further distinction may be made between the impact on oneself of what one may do wrong (sinning), the impact of the sins of other individuals or groups (being sinned against), and more indirect forms of suffering as a result of tyranny, oppression, and the ideological legitimation of unjust social orders (structural violence).[26] Both natural and social sources of suffering raise disturbing questions regarding God's care, but often in different ways. Accordingly, it seems that natural causes of suffering can only be attributed to God's punishment, while social causes of suffering would prompt protesting to God in God's name, given God's failure to care, allowing such suffering to endure. The delay of divine justice may be according to some vast eternal plan, but for the victims of history, that does not offer consolation. Again, such responses are amply attested to in the biblical witnesses, for example, in the frequent mentioning of "enemies" in the psalms.

It must be noted that the sources of suffering also include vicarious suffering—which is core to the Christian symbol of the cross. This raises soteriological questions of its own (e.g., regarding penal substitution), which will be addressed elsewhere in this series.[27] This does not resolve questions around natural suffering, though: is the vicarious role of the Savior necessary in order to patch up a "botched job" of the Creator?

However, a clear distinction between natural and social causes of suffering is not always possible, as is the case with psychosomatic diseases, the exacerbated impact of earthquakes, and pandemics such as human immunodeficiency virus (HIV) or coronavirus disease 2019 (COVID-19). The deeper reason for blurring the distinction between natural and social causes of suffering may well be the tendency to root evil in nature, for example, in the bodily passions, finiteness, materiality, temporality, primordial chaos, or, more typically, in anxiety over finitude. Evil then becomes a necessary function of finite existence (i.e., "metaphysical evil"). If so, it becomes possible to provide a rational explanation for the emergence of evil and therefore suffering. Ultimately, it is the Creator who is to be blamed! This prompts a tension between faith in God as Creator and as Savior. Often a deist notion of God is assumed in a critique of the Creator,

---

26. Such distinctions are widely discussed in the literature following Leibniz's classic distinction between moral, natural, and metaphysical evil. Among contributors to this series, see e.g. Conradie, *Redeeming Sin?*; also Moe-Lobeda, *Resisting Structural Evil*; Pearson, "Unwrapping Theodicy."

27. See Volume 5 of the series on *An Earthed Faith*, to be entitled "The Saving Grace of the Story?"

while some form of supernatural theism is assumed in a critique of the failure of God as Savior to rescue the victims of human history.

Nevertheless, the problem of natural sources of suffering remains, given that all suffering (e.g., animals inflicting pain on each other) cannot be attributed to a primeval human fall. Moreover, it seems that the very same evolutionary processes (at least partly through natural selection) that gave rise to beauty, diversity and complexity in nature also gave rise to suffering and extinction (or non-selection). Likewise, the very same natural processes that wreak havoc in human communities, including earthquakes, volcanic eruptions, tsunamis, and hurricanes, are necessary to maintain the immense fecundity of life on earth.[28]

Either way, believers are confronted with questions around God's loving presence and care amid multiple forms of suffering. In theological reflection, this has long prompted discourse on the intractable theodicy problem. All responses to the theodicy problem seem to remain deeply unsatisfying: "All grandiose theological systems that purport to have an answer to every question are exposed as illusory by the monstrous presence of evil and suffering in the world."[29] Some would conclude that any attempt to provide a rational explanation for God's presence amid suffering in history remains trapped in Hellenistic categories (such as λόγος, Εἱμαρμένη, πρόνοια), given their influence on early Christian notions of reason, the orderly succession of cause and effect, forethought, and determination. This suggests that evil things happen, as it were, not merely "under God's watch" but also "according to God's plan."

In response, others would insist that any human endeavor to find an explanation for (human) suffering, and a justification for God's failure to do something about that, will remain empty given God's response of forgiveness and the justification of the unjust.[30] Rational explanations for suffering typically amount to self-exoneration of human complicity. Yet others, like the character Ivan in Dostoyevsky's *The Brothers Karamazov*, would prefer to return their ticket for entry into God's kingdom if any unjust suffering is justified. This suggests that what is ultimately being questioned is God's character as loving and caring.

The blurring of distinctions between natural and social causes of suffering is epitomized by the "Anthropocene." This becomes evident in the distinction between ongoing changes in the Earth's atmosphere and anthropogenic climate change. The same applies to the distinction between

---

28. See Southgate, *Theology in a Suffering World*, 2.

29. See Migliore, *Faith Seeking Understanding*, 121.

30. See Berkouwer, *The Providence of God*, 260.

the concentration of sulfur in the Earth's atmosphere associated with volcanic eruptions and anthropogenic ocean acidification, or between the previous five mass extinctions and the current rapid loss of biodiversity due to a "sixth extinction." As the debates on naming the "Anthropocene" as such indicate, such anthropogenic sources of suffering at a planetary level are not homogenous but coincide with multiple social divides in terms of race, class, caste, and gender, and ideological divides in terms of economic systems, religious traditions, and worldviews.

■ Theological Reflection on Providence

In theological reflection on (human) suffering, it is necessary to clarify how God's providence relates to God's work of creation, ongoing creation, election, salvation, the formation of the church, God's mission, and the expected consummation of God's work.[31] There are at least four dangers here.

One is to separate God's care from, for example, God's work of salvation. That can only lead to theological speculation that becomes removed from the Christian gospel (a tendency that plagues intellectual attempts to resolve the theodicy problem) and the Christological hinge around which the Christian faith turns. Instead, some would insist that it is God's patience (μακροθυμίαν) to provide an opportunity for redemption that makes room for the story to continue (2 Pet 3:15). God's providence also cannot be reduced to the story of God's work of election (to be discussed in Volume 8 of the series) and salvation (to be discussed in Volume 5 of the series). This raises the thorny problem of how salvation history (a selection of events in the particular) and world history (in general) are related to each other. Clearly, neither fusion nor separation will do. Salvation history cannot be separated from the broad plane of (human) history, but it cannot be equated with that, either. History is one.[32]

The second danger is to subsume all these aspects of God's work under a single rubric. This leads to sterility, for example, where God's work becomes equated with the ministries and missions of the church. To subsume all of God's work under creation would lead to an arid deism or an amorphous pantheism. Instead, providence expresses the conviction that God, after creating the world, preserves and directs it to fulfill God's purposes.[33] To subsume providence (common grace) under salvation (salvific grace) runs the risk of understanding salvation as redemption

---

31. These themes will each be addressed in volumes 3 to 8 of the *An Earthed Faith* series.

32. Gutiérrez, *A Theology of Liberation*, 65.

33. See Fergusson, *The Providence of God*, 1.

from nature. To subsume providence under God's election of a covenant partner (Israel and the church) is to narrow down the scope of God's care for the whole of creation. Something similar would apply in other cases.[34] Clearly, providence by itself is not enough. More than providence is needed to respond to the "Anthropocene." But without providence as one dimension of God's work, all the others seem to unravel.

The third danger is to reduce reflection on God's providence to a discussion of the theodicy problem. The task may indeed be to expose flawed theodicies.[35] This may fill philosophical volumes but all too often tends to become arid and intellectualist and hardly ever suffices for a pastoral response to suffering.[36] In response, many contemporary theologians are drawn to the notion of tragedy to fathom the shadow side of God's good creation.[37] This prompts a related fourth danger, namely to shy away from the Jewish roots of the Christian notion of providence by adopting a Greek or even a Manichean notion of tragedy. This may either allow for a destructive force that is co-original with the Creator or for a shadow side to God's benevolent character.

How, then, should God's work of providence be understood and approached? There is simply no consensus in wider theological discourse in this regard. Often the doctrine of providence is itself sidelined as too complex and contested, even if this lies at the heart of a life of faith on a daily basis. For some, any notion of providence (e.g., through order in society) is far too much associated with the emphasis on development and progress in nineteenth-century liberal theologies or to the conservatism of a more orthodox theology that inhibits the possibility of a radically open future that would disrupt unjust structures in society. One may even say that there is little room for any doctrine of providence in several of the main theological trajectories of the twentieth century, including the early dialectical theology of the 1920s, the existential theologies of the 1950s, the political theologies of the 1960s and 1970s, or the subsequent

---

34. See Langdon Gilkey's critique that, for liberal theologies, providence provided the sole clue to the meaning of history; for dialectical theology it was Christology, while for political theologies it was eschatology. Clearly, these need to be juggled with each other in one or another story-line. See Gilkey, *Reaping the Whirlwind*, 238.

35. So Fergusson, *The Providence of God*, 303.

36. In a recent essay entitled "From Lisbon to Auschwitz and from Wuhan to Cape Town: COVID-19 as a Test Case for the Theodicy Problem" Ernst Conradie discusses a six-fold typology that he employs in teaching in this regard, under the following rubrics: (1) "Suffering is incomprehensible"; (2) "We need to offer protest before God about human suffering"; (3) "Human suffering is the product of human sin and may be regarded as God's punishment or chastisement for sin"; (4) "God is teaching us a lesson through suffering"; (5) "God's power is a function of persuasive love"; (6) "What is required is not passive acquiescence in suffering but courageous human participation in God's struggle against evil."

37. See e.g. Gandolfo, *The Power and Vulnerability of Love*.

emergence of a range of Black, decolonial, feminist, Indigenous, queer, and especially liberation theologies, including Dalit, Minjung, Pasifika, and other theologies.

In Protestant theology, a distinction is often made between three aspects of God's providence, namely *conservatio* (protection, sustenance), *gubernatio* (God's governance in history), and *concursus* (the interplay between divine, human, and other forms of agency, suggesting convivence and accompanying but not control, determinism, or domination).[38] A fourth aspect, namely *creatio continua* (ongoing creation), may be added, again with the danger of conflating creation and providence by subsuming both under ongoing creation.

These distinctions do provide some rubrics that may be employed, but these are not necessarily adopted in other confessional traditions. In the context of the "Anthropocene," the distinction between nature and (human) history has become blurred. Nature itself is historical, as the term natural history suggests, while humans not only form part of nature but have become a "geological force of nature." As a result, the distinction between God's work of conservation (in nature) and God's governance in history is also blurred.

To complicate matters further, the narrative account of planetary history in "history" as a human way of engaging with the past, as well as in history as an (academic) discipline, has become contested (note the triple meaning of the word "history").[39] The question is whether it is even possible to speak of history as such and whether any attempt to reconstruct such history does not fall into the trap of hegemonic grand narratives. There may be histories (his-stories and her-stories) in the plural, but any narrative account of such histories is already problematic, not to mention divine governance in history.

One may argue that conservation (maintaining order in nature) may well conflict with notions of governance in human history where social change is demanded instead of maintaining "law and order." Secular notions of governance and sovereignty are easily ascribed to God but then become contested given a critique of modern nation states. A notion of *concursus*

---

38. Here is the influential definition proposed by the Lutheran theologian Johann Andreas Quenstedt (1617-1688): "Providence is the external action of the entire Trinity whereby (1) God most efficaciously upholds the things created, both as an entirety and singly, both in species and in individuals; (2) Concurs in their actions and results; and (3) Freely and wisely governs all things to his own glory and the welfare and safety of the universe, and especially of the godly." Quoted in Wood, *The Question of Providence*, 78, note 10.

39. William Walsh identifies a double meaning of the ambiguous term "history," namely "the totality of human actions" and "the narrative or account we construct of them now." Note the anthropocentric focus on *human* history and the distinction between narrative accounts of history and academic reflections on such narrative accounts. See Walsh, *An Introduction to the Philosophy of History*, 16.

is required to avoid a deist, pantheist, or deterministic notion of *conservatio* and *gubernatio*, but it does not by itself express God's providence.[40] *Concursus* suggests that God does not act directly in history as an external cause but always through secondary causes, including human and other forms of agency. In the "Anthropocene," God's work of conservation in natural history can precisely no longer be separated from God's governance in human history.

■ Providence in Christian Ecotheology

In Christian ecotheology, such themes are widely addressed but not necessarily framed in terms of the doctrine of providence. At times, such themes become disconnected from a trust in God's care and are then reduced to a form of ethics, that is, in terms of human responsibility (which still assumes a position on *concursus*). Oddly, such ethics are often based (implicitly) on natural law, or God's creation ordinances (which played a notorious role in Nazi Germany and apartheid South Africa), or they amount to a theological appropriation of secular goals.

It may suffice here to list a number of discourses in Christian ecotheology (or touching on that) where such themes are indeed explored. Again, such themes are not necessarily framed in terms of providence, and any adequate treatment of such themes would need to do so not only in terms of providence. Nevertheless, they clearly do touch on an understanding of providence as well.

These discourses are placed here in a deliberately randomized order to illustrate how widely diverging and quite bewildering they are. References can be multiplied easily but are restricted here to a minimum.

Sustenance and Food Security

An expression of gratitude for sustenance and/or a cry for food amid hunger pangs are characteristic of daily piety, even though any prayer of gratitude for the food that we are about to receive is deeply troubled by unequal food distribution. Such piety emerges from the conviction that God cares. In ecotheology, there is by now a significant corpus of literature on food, food security, food sovereignty, eating, diets, Eucharistic practices, farming practices, concerns over the plight of farm animals, toxins,

---

40. The danger is that such cooperation between God and creatures can be reduced to either God using instruments in a way that would undermine their freedom, or God's work being regarded as a legitimation of human activity. While such cooperation is Christologically excluded (salvation for us), it is Pneumatologically necessary (salvation in and through us). See the famous essay by Arnold van Ruler on "Structural Differences between Christology and Pneumatology" in this regard.

monocultures, the use of biotechnology (genetically modified organisms), overfishing, and so forth. Such literature is typically found in the fields of Christian ethics, practical theology, and (Christian) spirituality but assumes some notion of God's providence. At the same time, the anthropocentrism that characterizes traditional discourse on God's providence (providing in human needs only) is being challenged.

Health and Healing

Alongside such an emphasis on food, there is an equally extensive theological discourse on health and healing, not least in the wake of the HIV and COVID-19 pandemics. The gendered nature of health is widely recognized, also but not only with reference to reproductive health, malnutrition, infant mortality, stunting in children, and debates around population and consumption. Especially in countries of the Global South, there is considerable debate regarding public health care systems, the role of pharmaceutical companies and inequalities in access to primary health care. Notions of health are readily extended to planetary well-being, as good health is impossible in a toxic environment.[41] This may be further extended to concerns over animal health, given the destruction of habitat, the plight of domesticated animals on commercial farms, and cruelty to animals. Again, such literature on health is found mainly in the fields of Christian ethics or pastoral care but assumes some notion of God's providence in giving us good health and strength, physically, psychologically, and socially.

Indigenous Wisdom

In Indigenous forms of ecotheology, the theme of God's providence may not feature as such, but the critique of imperialism, colonialism, and new forms of colonizing is required precisely because sustainable livelihoods are threatened. In response, Indigenous theologies retrieve ecological wisdom embedded in traditional forms of life, culture, and religion in order to survive amid forces of death and destruction.[42] At times, God's care is questioned, for example, by asking whether God is no longer adhering to the covenant with Noah not to destroy the world through water. Or to question, with Job, the sources of the suffering of the innocent. Or to find hospitable neighbors in cases of displacement (Luke 10).[43] In Indigenous

---

41. See Clinebell, *Ecotherapy*.

42. Among many examples worldwide, see e.g. Vaai, "The Ecorelational Story of the Cosmic Aiga." See also Vaai, "A Dirtified God."

43. See Talia, "Am I Not Your Tuakoi?"

communities, there is a growing interest in questioning and unsettling conventional views on God's providence as they struggle with colonial and neocolonial influences and the impact of anthropogenic climate change.[44] Indigenous communities are therefore not passively trusting that God's care will resolve such challenges but seek to reconstruct notions of providence. The resilience and courage of youth in Pacific islands is expressed in slogans such as "we are not drowning, we are fighting."[45] They are not passive, hopeless victims of climate change. Nevertheless, there is a sense of wonder and reverence for the fecundity of life that seems to become undermined in urbanized and industrialized contexts. It is precisely such reverence that is the source of moral courage to resist domination and exploitation.[46] On this basis, there is an interplay between divine and human agency (*concursus*).

Ecofeminist Critiques of Patriarchal Care

Any notion of God's providence is plagued by problems related to patriarchy, as God's care is traditionally associated with naming God as the Father of Jesus Christ. Accordingly, it is the task of the father in a patriarchal household to provide for the needs of all its members. Such problems are addressed in multiple forms of ecofeminist theology from around the world, as well as in LGBTQIA+ discourse. The challenge is aggravated by using the adjective "almighty." In a patriarchal context, an almighty father is downright scary! If God is male, then the male is God![47] This impression is reinforced through patriarchal church leadership. If God's care is expressed through priests and pastors as God's instruments ("pastoral care"), then sexual offences by priests and abusive charismatic leadership[48] (e.g., in demanding sexual favors, securing financial benefits, or occupying positions that carry social status) can only undermine any notion of God's providence. How, then, does one avoid a toxic masculinity in exploring the theme of God's providence? Can this be overcome merely by employing female images for God?

The Universe Story

The legacy of Pierre Teilhard de Chardin and Thomas Berry has become associated with reflections on the "universe story," especially the "journey

---

44. See Woodley, *Indigenous Theology and the Western Worldview*.

45. See Lusama, *Vaa Fesokotaki*.

46. See Chakrabarty, *The Climate of History in a Planetary Age*, 202.

47. See Daly, *Gyn/Ecology*.

48. See Herbert, "A Conceptual Analysis of Abusive Charismatic Leadership."

of the universe."[49] A sense of direction is discerned in cosmic, biological, and cultural evolution, arguably towards increasing diversity, complexity, autonomy, symbiotic networks, and beauty.[50] An ecological moral is discerned in this story by simultaneously upholding insights from contemporary science and retrieving wisdom from Indigenous communities and/or world religions.[51] Theological notions of God's governance in history may be deliberately eschewed, but the underlying question about the meaning of this moment in history (the "sixth extinction") remains obvious. Such an emphasis on the directionality of history is also employed to respond to a sense of cosmic meaninglessness (Stephen Weinberg),[52] given the laws of thermodynamics that may suggest that entropy will have the last work in the history of the universe. In response, a teleological understanding of the direction of history is affirmed by some, for example, in the process theology of John Haught.[53] God is not found primarily behind us in the Alpha as the First Cause but especially before us in the Omega, not in the beginning but at the end of the evolutionary process.[54]

■ Theologies of Becoming

There is a tendency to eschew the notion of providence and to subsume God's work of original creation and ongoing care under ongoing creation in theologies of becoming, for example, in the context of process theology.[55] Nevertheless, a strong emphasis on *concursus* (a classic theme discussed under the doctrine of providence) is found. There is often a stress on humans being created co-creators (Philip Hefner)[56] while recognizing some form of agency in each creature (Michael Welker).[57] Accordingly, God created things to create themselves. Not surprisingly, such an emphasis on ongoing creation can easily degenerate into either deism or pantheism.

---

49. See Swimme and Tucker, *Journey of the Universe*.

50. See also Rolston, *Genes, Genesis, and God; Three Big Bangs*.

51. See the critique of the assumptions on such ethics by Lisa Sideris in her *Consecrating Science*.

52. See the (in)famous quotation by Stephen Weinberg: "It is very hard to realize that this all is just a tiny part of an overwhelmingly hostile universe. It is even harder to realize that this present universe has evolved from unspeakably unfamiliar condition, and faces a future extinction of endless cold or intolerable heat. The more the universe seems comprehensible, the more it also seems pointless." Weinberg, *The First Three Minutes*, 154–55.

53. Such views are expressed throughout Haught's oeuvre. In the context of ecotheology, see already his *The Promise of Nature*.

54. See, for example, Ted Peters, *God—The World's Future*, following insights from Wolfhart Pannenberg.

55. See, especially, Keller, *Face of the Deep*.

56. See Hefner, *The Human Factor*.

57. See Welker, *Creation and Reality*.

If some form of theism is maintained, this raises the question of how God's care makes a difference in the world. Can God, for example, influence weather patterns to bring rain amid drought? Is merely praying for rain an appropriate response to climate change in the African context?[58]

Common Grace

Neo-Calvinist discourse on "common grace" is squarely located in the doctrine of providence, as the classic exposition by Abraham Kuyper also indicates.[59] The notion of common grace suggests that, in response to human sin, God decided not to destroy the work of God's hands. In order to implement God's plan for the salvation of the world (special grace) through divine election, God prevents the world from self-destruction in the interim through God's work of conservation and governance. Accordingly, God is "making room for the story to continue," keeping things from falling apart (W.B. Yeats, Chinua Achebe).[60] The emphasis on *common* grace is clearly inclusive and confirms God's benevolence to all people, indeed to the whole of creation. This is strengthened by the critique of imperial conquest (symbolized by the Tower of Babel)—the prime biblical reference used to support a notion of common grace. However, the very distinction between common and special grace tends to undermine such inclusivity. How common, then, is common grace? Moreover, the reference to creation ordinances as God's main tool for maintenance proved disastrous in Nazi Germany and apartheid South Africa (where such categories were explicitly adopted and adapted). Is "law and order" really what is required, or should we expect something radically new to move away from an oppressive past? But what, then, about the need for some continuity and an affirmation of the goodness of God's creation? Can one do away completely with some form of such ordinances or "mandates" (Dietrich Bonhoeffer)?[61]

Nature Conservation

Our human responsibility to engage in nature conservation and nature preservation is recognized especially among some groups of evangelicals but also in the Catholic and Orthodox monastic traditions. This may be in response to concerns over ecological destruction, soil erosion, the loss of

---

58. See Chitando and Conradie, "Praying for Rain?"

59. For a discussion, see the essay by Ernst Conradie in this volume.

60. Achebe, *Things Fall Apart*.

61. See Bonhoeffer's essay on the "The Concrete Commandment and the Divine Mandates" in his *Ethics*, 388–407.

biodiversity, species extinction, deforestation, and the like. This is typically captured through an emphasis on responsible stewardship and/or (in Orthodox theologies) an extension of human priesthood to address environmental concerns. The notion of stewardship is much criticized in some ecumenical circles, but the emphasis on responsibility is widely welcomed. There is ample literature on stewardship, and this can be quite nuanced, even if the metaphor as such is highly contested.[62] Here, the link between nature conservation and God's work of conservation and sustenance (as addressed under the rubric of God's providence) is obvious, if not always highlighted.

Sustainability[63]

There is, of course, a huge corpus of secular literature on sustainability, sustainable development, a sustainable society, and sustainable livelihoods. The term sustainability first gained global prominence, also in secular debates, at a conference on "The Future of Man and Society in a World of Science-Based Technology" hosted by the World Council of Churches (WCC) subunit on "Church and Society" in Bucharest (1974). The conference report acknowledges many impediments to hope but nevertheless expresses faith in God's providence that opens up new possibilities, even where these seem to be closed off by inequality, injustice, war, and environmental degradation.[64] Since that time, there have been many shifts and turns in secular discourse on sustainability, for example, moving from extending the use of nonrenewable resources ("Limits to growth"),[65] to the sustainable use of renewable resources (the Brundtland report),[66] to the carrying capacity of the land, to the absorption capacity of the atmosphere, to discourse on planetary boundaries. Theological discourse on sustainability is hardly ever framed in terms of the doctrine of providence, but it has obvious resonance with notions of *conservatio* understood as sustenance. Not surprisingly, the emphasis thus shifts from God's care to the human responsibility to ensure a sustainable future (see the notion of *concursus*). Yet, ultimately, one may say with Psalm 136 that it is God's mercy that sustains us forever.[67]

---

62. There is a huge corpus of ecumenical literature in this regard. For one critical overview, see Conradie, *Christianity and Earthkeeping*.

63. The discussion in this section draws on Conradie, "What, Exactly, Needs to Be Sustained amidst a Changing Climate?"

64. See the WCC's report on "Science and Technology for Human Development."

65. See Meadows et al., *Limits to Growth*.

66. See the Brundtland report entitled *Our Common Future*.

67. See Conradie, "Is It Not God's Mercy That Nourishes and Sustains Us … Forever?"

Natural "Evil"

Alongside discourse on structural evil in its many forms in conversation with the social sciences (see below), there is also considerable interest in natural sources of suffering in conversations between theology and the natural sciences, for example, in terms of natural catastrophes (often exacerbated by human factors), natural nonselection in biological evolution, sickness, fragility, degeneration, the role of predation and aggression among mammals, and so forth. Sometimes such natural sources of suffering are offered as an explanation for the emergence of human sin (through anxiety over finitude) and subsequently evil. As a result, the distinction between natural and social sources of suffering has become blurred. This, of course, raises the theodicy problem, namely to justify God's presence, sovereignty, and benevolence amid such overwhelming creaturely suffering.[68] Again, this is not necessarily linked with the theme of God's providence, but an affirmation of trust in God's care remains a response to the same underlying problem. In the "Anthropocene," as Clive Pearson observes, "[a]ll of a sudden, a change in the way the Earth System is understood lifts theodicy out of abstraction and places it within the hard realism of climate justice and possible endings."[69]

Structural Violence

In the wake of Auschwitz and Hiroshima, the focus of discussions of the theodicy problem shifted from natural causes of suffering (symbolized by the Lisbon earthquake)[70] to structural forms of evil and the ideologies that legitimize that.[71] Why does God allow evil to become so widespread? Can God not do something to overcome evil, to liberate the oppressed, to heal the wounds of the inflicted? Arguably, after 1945, at least in the West, symbolized by Jürgen Moltmann,[72] but also in the East, symbolized by Kazoh Kitamori,[73] theological discourse became dominated by the theodicy problem. In the Global South, the focus is typically on forms of structural evil (colonialism and neocolonialism) but not necessarily on the theodicy

---

68. In the context of ecotheology, see, especially, Southgate, *The Groaning of Creation*; *Theology in a Suffering World*.

69. Pearson, "Unwrapping Theodicy," 183. Pearson adds that, "For all the familiarity of the elements that make up this problem the term theodicy is seldom used in Oceania" (188).

70. For an extended comparison of Lisbon and Auschwitz as symbols of natural and social evil respectively, see Neiman, *Evil in Modern Thought*.

71. See, especially, Moe-Lobeda, *Resisting Structural Evil*.

72. This is marked by Jürgen Moltmann's *The Crucified God*.

73. See Kitamori, *The Theology of the Pain of God*.

problem. Instead, the focus is on social justice and perhaps questions on the relationship between God's justice and human struggles for liberation (*concursus*). Discourse on structural violence is readily extended to demands for ecojustice, given apocalyptic scenarios related to anthropogenic climate change, ocean acidification, and a rapid loss of biodiversity. Why does God not intervene to stop such catastrophes? Most theologians may wish to shy away from addressing this question, but it is not one that can be consistently avoided. There may be a wide range of responses, from prophetic resistance (in various liberation theologies) to apophatic unknowing, or (perhaps better) an interplay between these.[74]

Progression and Progress in History

In some secular debates, the nineteenth-century Western ideals of progress, based on the Enlightenment aspiration for rational control over "primitive chaos" and the presumed cultural sophistication (if not superiority) of Western civilization, remain prevalent if no longer as self-assured as they used to be. Such notions of progress emerged either in a naturalistic sense as the result of evolutionary forces in nature or in an idealistic sense as the unfolding of the Absolute Spirit (Hegel). Humans therefore play a decisive role in such evolutionary progress. Such notions of progress served as a functional replacement of Christian views of salvation so that Christianity itself could be readily replaced by demonstrable progress and increasing prosperity through the production of wealth and, given the role of labor unions, the distribution of such wealth. Following the social gospel paradigm, such progress could be associated with the coming of God's reign, bringing heaven down to earth. Alternatively, the nineteenth-century missionary zeal to proclaim the gospel to all nations to hasten in the kingdom served as an ecclesial version of such "progress." One finds this logic still embedded in linear notions of sustained economic growth, socio-economic development, the United Nations' sustainable development goals, the advance of modern science, technological innovation, advances in food production and medicine, and, more recently, the fourth industrial revolution. In ecumenical discourse, the growth paradigm is widely criticized, while notions of sustainability are questioned insofar as that assumes economic growth.[75] Nevertheless, there is a need to come to terms with advances in science and technological innovation, also in terms of artificial intelligence and ever more sophisticated military weapons.

---

74. For one recent example of such an apophatic theology, see Catherine Keller's *Political Theology of the Earth*.

75. See, for example, Béguin-Austin, *Sustainable Growth—Contradiction in Terms?* And for a recent overview, Conradie, "What, Exactly, Needs to be Sustained amidst a Changing Climate?"

Marxist and Other Utopias

Marxist critiques recognize the capitalist underbelly of such economic notions of progress but maintain an emphasis on the directionality of history, namely in the form of dialectic materialism towards the utopia of a classless society. Notions of the need, if not the necessity, for social transformation may be found in many other social movements, for example, towards emancipation from slavery and serfdom (against feudalism), women's emancipation (against patriarchal oppression), socio-economic liberation, decolonization, an inclusive democracy (against divisions of class and caste), an end to human trafficking, ending poverty and malnutrition, freedom of sexual expression (against heteronormativity), freedom of religious affiliation, and freedom from religious tutelage. In each case, such social transformation assumes that history moves or should move in a certain direction. Again, notions of God's providence are not necessarily referenced, but such movements also cannot be separated from God's governance in history.

Ecomodernist Optimism

One may also mention ecomodernist forms of optimism maintaining that the world is becoming a better place, if not for all humans and other animals, then at least on average, given the availability of food, disease control, a reduction of infant mortality, household technology, communication, longevity, and so forth. Some argue that there is a slow but still remarkable overall tendency to reduce violence through participatory decision-making, not only within countries but also globally (e.g., Steven Pinker).[76] There is also the contested thesis of Francis Fukuyama that history has reached its "end," given that no attractive and viable alternatives have emerged beyond a (capitalist?) market economy and a participatory (liberal?) democracy.[77] If there are indeed such alternatives (e.g., in the form of a "new economics"), then the burden of proof is on proponents to demonstrate their viability. At the same time, there are diverging models of a market economy and democracy so that further clarification is required. Ecumenical discourse on democracy and human rights, as well as on human and planetary well-being (intertwined as such themes are with notions of providence), cannot but be influenced by such secular debates, also where such optimism is subject to critiques from the perspective of the Global South.

---

76. See Pinker, *The Better Angels of Our Nature*.

77. See Fukuyama, *The End of History*.

Heading Towards a Catastrophe?

By contrast, there are also many prophets of doom that agree on the directionality of history but believe that industrialized civilizations are heading towards a catastrophe that can no longer be avoided.[78] If we are really doomed, what now?[79] Yet others place their hope in extraterrestrial explorations and astrobiology. Such varied secular assumptions are not necessarily legitimized in Christian theology, but insofar as these are not resisted, they may well become tacitly assumed. Christian eschatology can hardly be aligned with either an optimist or a pessimist teleology, but it does tend to envisage antinomies, contradictions, and tragic conflicts, without, however, succumbing to despair.[80] This should temper any feverish expectation for an ultimate triumph over evil. The question then remains how Christian eschatology relates to God's governance in history and therefore providence.

Discerning God's Finger in Human History

If such secular debates are necessarily reflected in theological discourses in one way or another, few are bold enough to discern God's finger in human and planetary history. In previous centuries notions of a "Holy Roman Empire," crusades *in hoc signo*, or the civilizing influence of the British Empire could be taken almost for granted in some contexts. An earlier generation was still able to reflect on the kingdom of God in America,[81] even on "*our* manifest destiny," or "God in South Africa,"[82] but most would refrain from any overt nationalist interests. There may be consensus that the Triune God is a God of history, but how God acts in history and whether God is steering history in some direction is regarded as riddled with so many controversies that only apophatic responses seem viable. Indeed, God's ways are not our ways (Isa 55:8) and cannot be easily traced (Rom 11:33-34). Clearly the presence of God's finger in history cannot be discerned merely from arbitrarily selected events, nor from extraordinary events only. Any event, including the exodus and the execution of Jesus, does not provide proof of God's providence by itself.[83]

---

78. See the earlier examples of Oswald Spengler's *The Decline of the West* and Arnold Toynbee's *A Study of History*, each with multiple editions.

79. See Scranton, *We're Doomed. Now What?*

80. See already Reinhold Niebuhr, *Faith and History*, 32.

81. See H. Richard Niebuhr, *The Kingdom of God in America*.

82. See Nolan, *God in South Africa*.

83. See Berkouwer, *The Providence of God*, 161-87.

It is only the Lamb that was slain who is able to open the book of history to discern evidence of God's reign (Rev 5:9).

The temptation to refrain from discerning God's presence in history leads almost inevitably to secularism or atheism. Other options include gnostic individualism or apocalyptic otherworldliness, but in both cases, God's care is no longer to be found in worldly affairs, in the flow of history.[84] With some Jewish scholars, there is a need to affirm that the God of Israel cannot be the God of past or future unless this God is still God of the present.[85] What, then, is this God up to amid the "Anthropocene"?[86]

A Theology of History

While there were ample contributions on God and world history in Western theologies up to the 1970s (e.g., Hendrikus Berkhof, Oscar Cullmann, Langdon Gilkey, Kornelis Miskotte, Jürgen Moltmann, Reinhold Niebuhr, Wolfhart Pannenberg, Paul Tillich, Arnold van Ruler, also Rubem Alves), few nowadays seem to dare to speak of the meaning of history. Some Western theologians prefer to avoid a theological interpretation of history altogether. The question of the meaning of history is deemed meaningless. Accordingly, history as such has no meaning, although we can *give* it meaning; we impose a meaning upon our constructions of history. Instead of considering the general direction of history, the focus shifts to the meaning of this moment (*kairos*) in history for me, perhaps for us, today (Tillich, Bonhoeffer). For others, the coming kingdom of God no longer arrives through a process of development on the basis of maintaining given orders in society but suddenly, through a disruption of unjust orders. The reign of God does not emerge through the steady influence of the past on the present but through the radical negation of the present by the promised future. With Rubem Alves, however, history is the medium through which God creates a future that is not yet there, breaking the present open towards the future.[87] In one way or another, therefore, Black, feminist/womanist, Indigenous, liberation, queer, postcolonial, and decolonial theologies further undermine universalist Western assumptions on history/his-story.

---

84. Fackenheim, *God's Presence in History*, 25. See also the comment by Arnold van Ruler: "Each time the Christian church has confessed to the salvific acts of God, this tended to result in a philosophy of history. Without this ferment, it would evaporate into idealism or Gnosticism or, currently, into existentialism or personalism [...] from a Christian perspective we *cannot* have God and Christ and salvation and truth without having history and its meaning." Van Ruler, *This Earthly Life Matters*, 180.

85. Rephrased from Fackenheim, *God's Presence in History*, 31.

86. See again Volume 2 of the *An Earthed Faith* series, entitled *How Would We Know What God Is Up To?*

87. See Alves, *A Theology of Human Hope*.

There nevertheless appears to be a market for popularized one-volume histories—of human evolution (Jared Diamond and many others), of humanity (Yuval Harari), of nearly everything (Bill Bryson), even of time itself (Stephen Hawking).[88] Either way, given the shift from the Holocene to the "Anthropocene," this task of interpreting the meaning and direction of history is one that cannot be avoided.

Providence and the Critique of Ideology

There is an obvious danger that the doctrine of creation may serve to legitimize the interests of the ruling class. As Vitor Westhelle has it, where landlords see the beauty of God's creation in looking over their estate, serfs and the landless only see gates and fences designed to keep them out.[89] The same applies to providence, as conservation tends to be understood in terms of God maintaining order in nature and society against the forces of chaos so that any oppressive "law and order" is easily legitimized as God's will. Likewise, God's governance in history is perceived by the victors of history, telling the story in such a way that reinforces constellations of power and privilege. Another example is the way in which "nature" is understood in the wake of the Enlightenment as an "object" to be scientifically dissected and objectively researched for the sake of human knowledge and using "natural resources" for the sake of economic growth. This remains a colonial and colonizing way of perceiving God's creation, where the one digests (the colon in *colon*ization) the available resources at the expense of others.[90] Indeed, belief in God's providence may well serve the purpose of acquiescence: "Providence is a tool invented by the rich to lull those whom they oppress into silent endurance. The rich have no need of virtue or faith for their desires are met without them. [...] Providence is either a tool invented for oppression or itself an instrument of injustice."[91] Should we, then, renounce the "god-trick" of a generalized, placeless, and disembodied vision of God's providence altogether?[92]

---

88. See Diamond, *The Third Chimpanzee*; Harari, *Sapiens*; Bryson, *A Short History of Nearly Everything*; Hawking, *A Brief History of Time*.

89. See Westhelle, "Creation Motifs in the Search for a Vital Space."

90. For a discussion of this etymological link between colon and colonization, see e.g. Vaai, "A Dirtified God."

91. Neiman, *Evil in Modern Thought*, 182.

92. See Clark and Szerszynski, *Planetary Social Thought*, 173.

Kenosis or Theosis?

Finally, two apparently contrasting views on the direction of history may be mentioned. The one finds a kenotic tendency in the history of the universe[93] that is exemplified in the incarnation, ministry, and death of Jesus Christ. This kenotic "principle" may also be read into "deep incarnation" (Niels Henrik Gregersen),[94] showing how suffering is built into and becomes fruitful in the history of biological evolution. The other expresses the typically Orthodox hope for divinization (theosis), for either becoming divine or for partaking in the life of the Triune communion. While the notion of kenosis perhaps has a rough secular equivalent in Stoicism, the hope for divinization is epitomized by the European Enlightenment and the subsequent critique of religion: God not only became human in Jesus Christ, but the death of god has to be proclaimed so that humans may become divine instead. This reading of history is exemplified in secular hopes for artificial, silicon-based forms of intelligence, for an Internet of all things, and indeed the hope to become divine as Homo excelsior or even Homo deus.[95] As Pope Francis astutely observes, "Artificial intelligence and the latest technological innovations start with the notion of a human being with no limits, whose abilities and possibilities can be infinitely expanded thanks to technology. In this way, the technocratic paradigm monstrously feeds upon itself."[96]

■ Challenges Posed by the "Anthropocene"

There is no need to resolve the obvious tensions between these notions of providence that are found not only but certainly also in Christian ecotheology. It may suffice to recognize these diverging notions of providence in order to add other possibilities and to ensure some form of ecumenical conversation across the many divides that characterize contemporary ecotheologies globally.[97]

What is at least clear is that the recognition of a shift from the Holocene to the "Anthropocene" has again raised questions about the linearity of history. Such linearity does not necessarily imply any progression or teleology, but things clearly do not always remain the same. Does *conservatio*

---

93. See Ellis and Murphy, *On the Moral Nature of the Universe*; also Rolston, "Kenosis and Nature."

94. See Gregersen, *Incarnation*.

95. See Harari, *Homo Deus*; Lynas, *The God Species*.

96. Pope Francis, *Laudate Deum*, §21.

97. In the context of contributions to this series, see, for example, Mendoza and Zachariah eds. *Decolonizing Ecotheology*.

mean that God will keep the earth in the Holocene epoch forever? Or should the focus be on God's governance, on a God of history who brings about an end to unjust social orders? What is it that has to end, and what may one hope could be sustained beyond the shift to the "Anthropocene"? How does one avoid imperialist and universalist claims in speaking of God's *gubernatio?* An apophatic response seems more appropriate, given the abuse of any notion of providence to legitimize the rise of industrialized capitalism, colonial conquest, and a consumerist society. There are moments when we ought not to speak. But then, of course, there are also times when we must speak.

Yet the challenges posed by the "Anthropocene" to Christian (eco) theology remain. Ever since the rise of Greek philosophy, there has been an attempt to demythologize history, to clear it from Homeric gods. Contemporary Jewish and Christian theologians seem to follow suit. Just as God needs to be expelled from nature by modern scientists, it seems that God also needs to be expelled from history.[98] But in the "Anthropocene," a clear distinction between nature (God's *conservatio*) and (human) history (God's *gubernatio*) can no longer be maintained, as the impact of some human cultures is now inscribed in the atmosphere for hundreds of thousands of years to come and in the Earth's rock layers for millions of years to come.

How could any retrieval of God's providence respond to that? There seems to be relatively few options available. One is to (plead for a) return to a pre-Jewish cyclical view of history by retrieving Indigenous notions of wisdom. If divine providence and the evils of history prove incompatible, is it not best to disconnect God from history altogether? Or perhaps history follows a completely random path in which no pattern or divine presence can be discerned? Another is to boldly adopt some form of teleology, finding something like the kingdom of God at the "end" of history. Any such a notion of progression towards a final goal always comes at the risk of underplaying the impact of human sin, to which God's providence is supposedly a response. A third is a gnostic escape from history in order to find what abides despite historical fluctuations. There is also an apophatic option to live with current contradictions, finding God *sub contrario*. Is this an escape mechanism as well? What, then, about the hope for the consummation of God's work? From the perspective of faith in God's providence, the least one may say, following Rowan Williams, is that "God's faithfulness stands, assuring us that even in the most appalling disaster, love will not let go."[99]

---

98. See Fackenheim, *God's Presence in History*, 4–5.

99. Rowan Williams, *Faith in the Public Square*, 190.

Perhaps the core intuition of the Christian notion of providence is to linger a bit longer in God's work of *conservatio*, *gubernatio*, and *concursus*, before bringing God's work of salvation, the church and its ministries and missions, and God's work of consummation into play too quickly. Admittedly, each of these themes cannot be addressed adequately in terms of the doctrine of providence alone. God's governance, for example, is best discussed with reference to the reign of God, Christ's sitting "at the right hand of the Father"[100] and the role of the Paraclete in the rule of law. Likewise, any adequate theology of history cannot be reduced to God's governance in terms of the doctrine of providence but would require reflection on the Christological dialectic between cross and resurrection and on the Pneumatological relatedness of proton, history, and eschatological consummation.

Nevertheless, providence cannot be subsumed under such other categories either. It has a place of its own, as contested as that may be. In other words, there may be a need to make a bit of room for the story to continue …

■ Bibliography

Alves, Rubem. *A Theology of Human Hope*. St. Meinrad: Abbey Press, 1969.

Barth, Karl. *Church Dogmatics III.3*. Edinburgh: T&T Clark, 1960.

Béguin-Austin, Midge, ed. *Sustainable Growth—Contradiction in Terms?* Geneva: The Ecumenical Institute, 1993.

Berkhof, Hendrikus. *Christ the Meaning of History*. Grand Rapids: Baker, 1965.

Berkouwer, Gerrit C. *The Providence of God*. Grand Rapids: Eerdmans, 1952.

Bonhoeffer, Dietrich. *Ethics. Dietrich Bonhoeffer Works 5*. Minneapolis: Fortress Press, 2005.

Brundtland, Gro Harlem, ed. *Report of the World Commission on Environment and Development: Our Common Future*. 1987. https://sustainabledevelopment.un.org/content/documents/5987our-common-future.pdf.

Bryson, Bill. *A Short History of Nearly Everything*. New York: Broadway, 2003.

Chakrabarty, Dipesh. *The Climate of History in a Planetary Age*. Chicago: University of Chicago Press, 2021.

Chitando, Ezra, and Ernst M. Conradie, eds. "Praying for Rain? African Perspectives on Religion and Climate Change." *The Ecumenical Review* 69:3 (2017) 311–435. https://doi.org/10.1111/erev.12304.

Clark, Nigel, and Bronislaw Szerszynski. *Planetary Social Thought: The Anthropocene Challenge to the Social Sciences*. Cambridge: Polity, 2021.

---

[100]. This Christological emphasis on Christ sitting at the right hand of the Father (in the present tense) is, as may be expected, a core affirmation in Karl Barth's treatment of the *Christian* doctrine of providence. He also points to Colossians 1:17, where it is said that in Christ all things hold together (συνέστηκεν) and to Hebrews 1:3 affirming that all things are upheld (φέρων) by the powerful word of the Son sitting at the right hand of the Majesty on high. See Barth, *Church Dogmatics III.3*, 35. For Barth, God as Father is not a stranger, alien or an enemy of the creature, or the subject of a father-complex, but the Father of Jesus Christ and therefore *our* beloved Father (146).

Clinebell, Howard J. *Ecotherapy: Healing Ourselves, Healing the Earth*. New York: Haworth Press, 1996.

Conradie, Ernst M. *Christianity and a Critique of Consumerism: A Survey of Six Points of Entry*. Wellington: Bible Media, 2009.

——. *Christianity and Earthkeeping: In Search of an Inspiring Vision*. Stellenbosch: SUN, 2011.

——. "From Lisbon to Auschwitz and from Wuhan to Cape Town: COVID-19 as a Test Case for the Theodicy Problem." In *World Christianity and COVID-19: Looking Back and Looking Forward*, edited by Chammah Kaunda, 19–34. Berlin: Springer Nature, 2023.

——. "Is It Not God's Mercy That Nourishes and Sustains Us … Forever? Some Theological Perspectives on Entangled Sustainabilities." *Scriptura* 116 (2017) 38–54. https://doi.org/10.7833/116-2-1330

——. *Redeeming Sin? Social Diagnostics and Ecological Destruction*. Langham: Lexington, 2017.

——. "Some Theological Reflections on Multi-disciplinary Discourse on the 'Anthropocene'." *Scriptura* 121 (2022) 1–23. https://doi.org/10.7833/121-1-2076

——. "What, Exactly, Needs to be Sustained Amidst a Changing Climate?" In: *Global Sustainability. Issues in Science and Religion*, edited by Michael Fuller et al., 25-40. Cham: Springer, 2023. https://doi.org/10.1007/978-3-031-41800-6_3.

Conradie, Ernst M., with Willie James Jennings. "The Place of Creation in Christian Ecotheology—Some Shifts in the Story." In *The Place of Story and the Story of Place*, edited by Ernst M. Conradie and Willie James Jennings, 1–24. Cape Town: AOSIS, 2024.

Conradie, Ernst M., with Pan-Chiu Lai. "On Setting the Scene for the Story to Begin." In *Taking a Deep Breath for the Story to Begin … An Earthed Faith: Telling the Story Amid the "Anthropocene" Volume 1*, edited by Ernst M. Conradie and Pan-Chiu Lai, 15–48. Cape Town: AOSIS, 2021.

Conradie, Ernst M., and Cynthia D. Moe-Lobeda. "Telling the Story *en route*: On This Road (*hodos*) and its Logic (*logos*)." In *How Would We Know What God Is Up To? An Earthed Faith Volume 2*, edited by Ernst M. Conradie and Cynthia Moe-Lobeda, 1–30. Cape Town: AOSIS, 2022.

Daly, Mary. *Gyn/Ecology: The Metaethics of Radical Feminism*. Boston: Beacon, 1978.

Diamond, Jared. *The Third Chimpanzee: The Evolution and Future of the Human Animal*. New York: HarperCollins, 1991.

Durand, J. J. F. (Jaap). *Skepping, Mens, Voorsienigheid*. Pretoria: N.G. Kerkboekhandel, 1982.

Ellis, George F. R., and Nancey Murphy. *On the Moral Nature of the Universe*. Minneapolis: Fortress Press, 1996.

Fackenheim, Emil. *God's Presence in History: Jewish Affirmations and Philosophical Reflections*. New York: Harper Torchbooks, 1970.

Fergusson, David. *Creation*. Grand Rapids: Eerdmans, 2014.

——. *The Providence of God: A Polyphonic Approach*. Cambridge: Cambridge University Press, 2018.

Francis (Pope). *Laudate Deum: Apostolic Exhortation of the Holy Father Francis to All People of Good Will on the Climate Crisis*, given at the Basilica of Saint John Lateran on the 4th of October on the Feast of Saint Francis of Assisi. https://www.vatican.va/content/francesco/en/apost_exhortations/documents/20231004-laudate-deum.pdf.

Fukuyama, Francis. *The End of History and the Last Man*. New York: Free Press, 1992.

Gandolfo, Elizabeth O'Donnell. *The Power and Vulnerability of Love: A Theological Anthropology*. Minneapolis: Fortress Press, 2015.

Gilkey, Langdon. *Reaping the Whirlwind: A Christian Interpretation of History*. New York: Seabury, 1976.

Gorringe, Timothy. *God's Theatre: A Theology of Providence*. London: SCM, 1991.

Gregersen, Niels Henrik, ed. *Incarnation: On the Scope and Depth of Christology*. Minneapolis: Fortress, 2015.

Gutiérrez, Gustavo. *A Theology of Liberation*. Maryknoll: Orbis, 1973.

Hamilton, Clive. "Anthropocene as Rupture." *Anthropocene Review* 3 (2016) 93–106. https://doi.org/10.1177/2053019616634741

Harari, Yuval Noah. *Homo Deus: A Brief History of Tomorrow*. London: Vintage, 2017.

———. *Sapiens: A Brief History of Humankind*. New York: Hatchett Books, 2020.

Haught, John F. *The Promise of Nature: Ecology and Cosmic Purpose*. Mahwah: Paulist, 1993.

Hawking, Stephen. *A Brief History of Time: From the Big Bang to Black Holes*. New York: Bantam, 1988.

Hefner, Philip. *The Human Factor: Evolution, Culture and Religion*. Minneapolis: Fortress, 1993.

Herbert, Brian. "A Conceptual Analysis of Abusive Charismatic Leadership." MPhil mini-thesis, University of the Western Cape, 2008.

Keller, Catherine. *Face of the Deep: A Theology of Becoming*. New York: Routledge, 2003.

———. *Political Theology of the Earth: Our Planetary Emergency and the Struggle for a New Public*. New York: Columbia University Press, 2018.

Kitamori, Kazoh. *The Theology of the Pain of God*. Eugene: Wipf & Stock, 2005.

Lusama, Tafue. *Vaa Fesokotaki: A Mafulifuli Reconstruction of the Theology of te Atua for a New Tuvalu Climate Change Story*. Suva: Pacific Theological College Press, 2022.

Lynas, Mark. *The God Species: Saving the Planet in the Age of Humans*. Washington, DC: National Geographic Society, 2011.

Meadows, Donella H., et al. *The Limits to Growth: A Report for the Club of Rome's Project on the Predicament of Mankind*. New York: Universe Books, 1972.

Mendoza, S. Lily, and George Zachariah, eds. *Decolonizing Ecotheology: Indigenous and Subaltern Challenges*. Eugene: Pickwick, 2021.

Migliore, Daniel L. *Faith Seeking Understanding: An Introduction to Christian Theology*. 3rd ed. Grand Rapids: Eerdmans, 2014.

Moe-Lobeda, Cynthia D. *Resisting Structural Evil: Love as Ecological-Economic Vocation*. Minneapolis: Fortress, 2013.

Moltmann, Jürgen. *The Crucified God*. New York: Harper, 1974.

Moore, Jason W., ed. *Anthropocene or Capitalocene? Nature, History and the Crisis of Capitalism*. Oakland: PM Press, 2016.

Neiman, Susan. *Evil in Modern Thought: An Alternative History of Philosophy*. Princeton: Princeton University Press, 2002.

Niebuhr, H. Richard. *The Kingdom of God in America*. New York: Harper & Brothers, 1937.

Niebuhr, Reinhold. *Faith and History: A Comparison of Christian and Modern Views of History*. London: Nisbet & Co., 1949.

Nolan, Albert. *God in South Africa: The Challenge of the Gospel*. Cape Town: David Philip, 1988.

Pearson, Clive R. "Unwrapping Theodicy." In *Theologies from the Pacific*, edited by Jione Havea, 181–92. New York: Palgrave Macmillan, 2021.

Peters, Ted. *God—The World's Future: Systematic Theology for a New Era*. 3rd ed. Minneapolis: Fortress, 2015.

Pinker, Steven. *The Better Angels of Our Nature: The Decline of Violence in History and its Causes*. New York: Penguin, 2011.

Rieger, Joerg. *Theology in the Capitalocene: Ecology, Identity, Class, and Solidarity*. Minneapolis: Fortress, 2022.

Rolston, Holmes, III. *Genes, Genesis and God: Values and their Origins in Natural and Human History*. London: Cambridge University Press, 1999.

———. "Kenosis and Nature." In *The Work of Love*, edited by John Polkinghorne, 43–65. Grand Rapids: Eerdmans, 2001.

———. *Three Big Bangs: Matter-Energy, Life, Mind*. New York: Columbia University Press, 2010.

Scranton, Roy. *We're Doomed. Now What? Essays on War and Climate Change*. New York: Soho Press, 2018.

Sideris, Lisa H. *Consecrating Science: Wonder, Knowledge and the Natural World*. Oakland: University of California Press, 2017.

Southgate, Christopher B. *The Groaning of Creation: God, Evolution, and the Problem of Evil*. Louisville: Westminster John Knox, 2008.

———. *Theology in a Suffering World: Glory and Longing*. Cambridge: Cambridge University Press, 2018.

Swimme, Brian, and Mary Evelyn Tucker. *Journey of the Universe*. New Haven: Yale University, 2011.

Talia, Maina. "Am I Not Your Tuakoi? A Tuvaluan Plea for Survival in Time of Climate Emergency." DPhil thesis, Charles Sturt University, 2023.

Vaai, Upolu Lumā. "A Dirtified God: A Dirt Theology from the Pacific Dirt Communities." In *Theologies from the Pacific*, edited by Jione Havea, 15–30. New York: Palgrave Macmillan, 2021.

———. "The Ecorelational Story of the Cosmic Aiga: A Pasifika Perspective." In *Taking a Deep Breath for the Story to Begin ... An Earthed Faith: Telling the Story Amid the "Anthropocene" Volume 1*, edited by Ernst M. Conradie and Pan-Chiu Lai, 225–40. Cape Town: AOSIS, 2021.

Van Ruler, Arnold A. *This Earthly Life Matters: The Promise of Arnold van Ruler for Ecotheology*, edited by Ernst M. Conradie. Eugene: Wipf & Stock, 2023.

———. "Structural Differences between Christology and Pneumatology." In *Calvinist Trinitarianism and Theocentric Politics*, translated and edited by John Bolt, 27–46. Lewiston: Edwin Mellen, 1989.

Walsh, William H. *An Introduction to the Philosophy of History*. London: Hutchinson & Co, 1967.

Weinberg, Stephen. *The First Three Minutes*. New York: Basic Books, 1977.

Welker, Michael, *Creation and Reality*. Minneapolis: Augsburg Fortress, 1999.

Westhelle, Vitor. "Creation Motifs in the Search for a Vital Space: A Latin American Perspective." In *Lift Every Voice: Constructing Christian Theologies from the Underside*, edited by Susan Brooks Thistlethwaite and Mary Engel Potter, 146–58. Maryknoll: Orbis, 1998.

Williams, Rowan. *Faith in the Public Square*. London: Bloomsbury, 2012.

Wood, Charles. *The Question of Providence*. Louisville: Westminster John Knox, 2008.

Woodley, Randy. *Indigenous Theology and the Western Worldview: A Decolonized Approach to Christian Doctrine*. Grand Rapids: Baker Academic, 2022.

World Council of Churches, Sub-Unit on Church and Society. "Science and Technology for Human Development: The Ambiguous Future and the Christian Hope." *Anticipation* 19 (1974) 2–43.

Divine Providence amid Ecological Crises: A Nigerian Perspective

Aku S. Antombikums[1]

Introduction

While growing up in a small village in Nigeria, there was a forest adjacent to the village covering over a hundred hectares of land. After high school, I went to the city and, after some time, decided to travel back to the village. I was shocked to find out that the forest was gone. I was told that it was the new village chief, who had resigned from the priestly ministry, who ordered that the whole forest be ransacked because it harbored evil spirits, as it used to be the village shrine.

Ecological crises (broadly construed) raise existential worries similar to the moral or social problem of evil. One such existential worry is the main focus of this volume: "How could the suffering of God's creatures in the

---

1. Aku Stephen Antombikums (PhD) is a research associate at the University of Pretoria, South Africa and a research fellow at the Theological University, Utrecht, the Netherlands.

How to cite: Antombikums, AS 2024, 'Divine Providence amid Ecological Crises: A Nigerian Perspective', in EM Conradie & UL Vaai (eds.), *Making Room for the Story to Continue?*, An Earthed Faith: Telling the Story amid the "Anthropocene", vol. 4, AOSIS Books, Cape Town, pp. 35-53. https://doi.org/10.4102/aosis.2024.BK415.02

'Anthropocene' be reconciled with trust in God's loving care?" As a result of this question, the nature of God's governance also comes into play, including God's goodness and power. Christian theologians and philosophers agree that God is the Lord of creation. Arguably, because God is the Lord of creation, God is assumed to control all that happens in the universe. However, to what extent God is controlling everything in the universe seems debatable. Is God controlling everything or only some things? How do we reconcile God's goodness and divine control with current ecological woes? Because the doctrine of divine providence revolves not only around divine governance but also around divine conservation and *concursus*, what notion of divine providence is sufficiently robust for understanding and tackling contemporary ecological crises?

Classical theists (Augustine and Calvin, for instance) argue that God exercises "meticulous providence"[2] over creation, including cases of natural disaster. Contrary to classical theism, open theists argue that God exercises general providence, given that the creation is an open project with open routes. In this essay, I will investigate the influences of these two opinions on divine control in relation to the climate crisis. Which view is viable in dealing with the ecological crisis? In other words, which of the two is most promising when it comes to encouraging us to combat climate change and other ecological threats? Would classical theism be more promising, given that it teaches that God is fully in control and therefore assures us that there is no need to "freeze" out of total despair (as some non-Christians do)? Or would open theism be preferred as it takes human action more seriously, not suggesting that everything is predetermined by God's will or foreknowledge already?

Because the essays by Ernst Conradie and Gijsbert van den Brink in this volume deal with the doctrine of divine providence, I will avoid repeating such issues discussed in their papers and elsewhere in this volume. In what follows, I will introduce classical theism and open theism briefly to show how such conceptions have influenced Christians in Africa, who have weaved such notions with a traditional pre-Christian worldview to come up with a cultural brand of Christianity that is quite distinctive. For example, most Christians in Nigeria do not subscribe to either classical theism or open theism in the Wesleyan tradition. Some Nigerian Christians argue that the current climate crisis must be understood eschatologically. Unfortunately, others encourage deforestation by believing that evil spirits inhabit such a forest. Taking classical theism seriously on the one hand and open theism on the other, I will show the tensions associated with these

---

2. For the use of this term, see Crisp and Sanders's *Divine Action and Providence: Explorations in Constructive Dogmatics*.

two views in dealing with the climate crisis and how such tensions have impeded viable responses to the ecological crisis, especially from within the Global South, in order to search for adequate responses to climate crisis, notwithstanding theological differences.

■ Providence Revisited[3]

For many ages in human history, until the Enlightenment period, everything was considered to have happened by divine arrangement. However, with the passage of time, the notion of divine providence was rejected for several reasons.[4] Although this rejection was not the origin of the "Anthropocene," it contributed to industrialization and cultural revolutions, emphasizing the human person and her needs and giving rise to the notion of self. Traditions that maintain hierarchical boundaries have also collapsed. And so today, we live in a fluid world. This loss of divine providence is evident from the advent of the "Anthropocene" as a geological period following the Holocene.

To offer an account for the seeming loss of a concept of divine providence in human history, David Fergusson asserts that two significant changes occurred in church history which challenged the notion that everything is governed by divine providence. The first is a rejection of the supernatural. When rationalists began to advocate for a general providence linked to the laws of nature rather than to a divine being, they were laying the foundation for the rejection of divine providence. The rejection of miracles by David Hume is one such example.[5] Although not everyone rejected the miracles recorded in the Bible as Hume did, the church tended to confine miracles to the apostolic era.[6] In their insistence that miracles ended during the apostolic era of the church, such providentialists had begun championing a different notion of divine control, which had precursory implications for how divine providence, especially in the "Anthropocene," is conceived.[7]

Given that miracles were confined to the apostolic era, including the rejection of meticulous providence, the traditional notion of divine

---

3. This section, including the sections on classical and open theisms are drawn and reworked from Aku S. Antombikums, 2022, 'Open theism and the problem of evil', PhD thesis, Vrije Universiteit Amsterdam, the Netherlands, supervised by Prof. Dr Henk van den Belt and Prof. Dr Rik Peels, viewed 26 September 2024, at https://research.vu.nl/ws/files/171313479/A%20S%20%20Antombikums%20-%20thesis_redacted.pdf?

4. Genevieve Lloyd argues that "Providence may now be largely 'lost' from our secular consciousness; but it continues to exert an influence on our thought and on our lives." See her *Providence Lost*, 1. Cf. Antombikums, *Open Theism and the Problem of Evil*, 17.

5. See Chapter X of David Hume's *An Enquiry Concerning Human Understanding*.

6. Fergusson, *The Providence of God*, 111.

7. See Antombikums, *Divine Control* (May 2025 forthcoming).

providence was also rejected from a deist perspective[8] because the responses that the church provided for natural evil, especially the Lisbon earthquake, were deemed unsatisfactory: "The Lisbon disaster was a hinge event for many thinkers, particularly Voltaire, who attacked the optimism of Leibniz and Pope's insistence that 'all is well' [...] Voltaire also mocked the notion that this could contribute to some greater good or to the achievement of a collective whole."[9]

Today, in the words of the apostle, the creation "groans" excessively as a result of human activities, beginning with the period of the industrial revolution, weapons of mass destruction as seen in the bombing of Hiroshima and Nagasaki, constant acts of terrorism, and the destabilization of ecosystems through human activities like desertification, ocean acidification, water pollution, ozone depletion, and food wastage.

The rupturing of the ecosystem in the "Anthropocene" raises many questions with respect to a theistic concept of divine providence. In other words, is the doctrine of divine providence still plausible today? Should we believe that God is still active in the "Anthropocene"? How do we uphold the idea of God's love and care amid the current ecological crisis? In what follows, I will summarize classical theism and open theism (as an offshoot of Wesleyan theology, sharing some connections with Pentecostalism) and state the obvious problems arising from the appropriation of these two notions of divine providence. In my view, an essential issue here is the question of whether God macro-manages or micro-manages the universe and what implications this may hold for addressing ecological destruction.

■ Classical Theism

The debate about divine providence, including what should be an appropriate religious response to the ecological crisis, seems to originate in the doctrine of creation. To this end, classical theists situated their doctrine of divine providence in the doctrine of creation. For instance, in Augustine's thought, the notion that God creates everything *ex nihilo*[10] gives God the right *prima facie* to exercise every form of divine governance over the cosmos. In this regard, William Mann argues that "God's sovereignty over all other things is grounded in the fact that he created them."[11] As the

---

8. Fergusson, *The Providence of God*, 115. Cf. Antombikums, *Open Theism and the Problem of Evil*, 18.

9. Fergusson, *The Providence of God*, 124. Cf. Antombikums, *Open Theism and the Problem of Evil*, 18.

10. Open theists are divided on the notion of *creatio ex nihilo*. For instance, John Sanders, Clack Pinnock, and William Hasker support this idea, while Thomas Oord finds it problematic.

11. Mann, "Augustine on Evil and Original Sin," 42–43.

Creator of the cosmos, God alone can cause the creation to exist in a state of quietness through providence.

Contrary to the accusation leveled against the classical doctrine of providence as tilting towards fatalism, Mark Elliot argues that "Augustine sees providence as working in two ways: the natural operation and God's voluntary operation, and, as that affects humankind, for the sake of the body and then of the soul. Both are equally valued by God."[12] This conclusion seems to be close to the medieval distinction between primary and secondary causes of actions championed by Thomas Aquinas and later mentioned by Calvin. In response to his critics, Augustine argues that "[a]nyone who doubts the existence of this divine Providence, for the sake of coherence, will also have to admit the irrationality of the creative action of God and recognize that God, both at the moment of creation and previously, did not know what He was doing because He would have been deprived of reason, which is the criterion of his work."[13] Contrarily, the divine finger must be seen in how the law of nature functions, including historical happenings. In divine providence, God's active power for every individual creature is at play.[14]

David Fergusson argues that just as Augustine sees the operation of divine providence in the law of nature, including every historical event, so does Calvin, who sees God's providence as encompassing inanimate objects, microorganisms, and humans.[15] Although *prima facie* the conclusion above, when overstretched, could lead to fatalism,[16] Paul Helm thinks this is not the case. He argues that "Calvin is able to distinguish between (a) human purpose and intention and (b) the higher purpose and intention of God in ordaining the action of the secondary cause."[17]

Calvin argues that the natural order, including inanimate objects and human actions, is meticulously ordered by God's "secret" decree so that every state of affairs happens because God willingly and knowingly desired that it should happen in that way. As far as Calvin is concerned, there is no such thing as general providence.[18] As a result, even "[t]hose things which are vainly or unrighteously done by man are, rightly and righteously, the

---

12. Elliot, *Providence Perceived*, 39.

13. Augustine, *The Retractations* 1.3.2.

14. Fergusson, *The Providence of God*, 84.

15. Fergusson, *The Providence of God*, 86-88.

16. Muller, *Divine Will and Human Choice*, 187-89.

17. Helm, *The Providence of God*, 179.

18. Calvin, *Institutes of the Christian Religion*, 1.16.3, 4, 5, 6, 7.

works of God!"[19] Accordingly, the glory of God is more significant to God than everything else, to the extent that Calvin is willing to accept that God ordained all human actions, including their sins, instead of accepting that there are things that are not under divine control.

■ Wesleyan/Open Theistic View of Divine Control

Contrary to the position of classical theism, as sketched above, the Wesleyan tradition has always emphasized human freedom in the context of salvation.[20] Today, open theism, an offshoot of the Wesleyan tradition, maintains that the classical notion of meticulous or exhaustive providence renders human freedom useless. It logically implies that God is the author of all human actions, whether good or bad. For open theism, creation suggests the contrary of the classical notion of divine providence. Creation is not an event that happened once in the past;[21] it is an ongoing process with open routes and without a specific goal.[22]

In empowering humans to care for creation, God also grants them the power to bring into existence certain states of affairs that were not in existence at the dawn of creation. Occupying the land and multiplying, as commanded in Genesis 1:28, means that the land "[…] will take on characteristics it did not have on the seventh day."[23] Although John Sanders does not spell out the implications of this conclusion for ecological challenges, it is evident that such new characteristics may include ecological concerns. I will return to this shortly. Although classical theists developed a meticulous view of divine providence from the Hebrew Bible, open theists argue that, by choosing to empower the creation instead of creating them without such powers, despite being sovereign, God has voluntarily chosen to exercise a general providence rather than meticulous providence. This therefore means that God is a risk-taker.[24]

Contrary to the notion of *creatio ex nihilo* upheld in classical theism, which gives God the right *prima facie* to exercise meticulous providence over creation, Thomas Oord, an open theist, advocates for a notion of being co-creators with God. He argues that the notion of being co-creators could

---

19. See Calvin, *A Defence of The Secret Providence of God*.

20. See Arminius, *The Works of James Arminius* (3 Volumes).

21. Sanders, "Open Creation and the Redemption of the Environment," 141, 142–43, 146.

22. Rice, *God's Foreknowledge and Man's Free Will*, 37.

23. Sanders, "Open Creation," 142.

24. Sanders, *The God Who Risks*, 40.

provide two viable grounds for responding to climate change: (a) Because God did not create everything single-handedly, the current state of the world, whether good or bad, is not to be understood from a divine action point of view alone but a combination of divine and human actions. In that case, humans may be praised or blamed for the climate crisis. (b) This provides the theoretical basis for a viable approach to halting the climate crisis. It simply means that humans need to change their habits that aggravate the ecological crisis, including enacting the right laws to combat it.[25]

Pentecostalism may be regarded as a significant contemporary wing of the Wesleyan tradition. It has been observed that about 35 percent of Christians (about 203 million) in Africa are Pentecostals.[26] As I have argued elsewhere, "Pentecostalism [...] shares a couple of things in common with open theism, which is a branch of freewill theism. This includes the emphasis on freedom in relation to salvation, the risk-taker model of God and its implications for petitionary prayers, and the role of experience in theological and biblical interpretation."[27] Its emphasis on freedom, prosperity, dominion, and the like on the one hand and upholding strong determinism on the other hand could be both promising and discouraging in combating ecological problems. Pentecostal churches, especially in Nigeria, engage in what could be termed "the African primal quest for the sacred and the transcended: the quest for healing, well-being, material success, and long life."[28]

Another common feature between African Pentecostalism and open theism is that both are rooted in American Pentecostalism—which also maintains vital elements of Wesleyan theology, as mentioned above. Although Nigerian Pentecostal churches are deeply rooted in African Independent Churches (AICs), they do not concur with AICs in all respects: "For example, African prophetic churches and Pentecostal-charismatic churches, while both condemning African ritual practices such as divination, ancestor veneration, traditional medicine, and healing, paradoxically share other aspects of indigenous orientation, such as visions, dreams, healing, 'spirit' possession, and divine revelation."[29]

In my assessment, the difference between Pentecostalism and AICs is that the latter remain cultic and do not seem to exhibit a life of affluence.

---

25. Oord, "God's Initial and Ongoing Creating," 267.

26. Wariboko, "Pentecostalism in Africa," 5.

27. Antombikums, "African Pentecostalism and its Dilemma," 6.

28. Olupona, "West Africa," 12.

29. Olupona, "West Africa," 12.

Charismatic-Pentecostal churches not only weaved Wesleyanism into their theology but also endorsed the self-gratification and spontaneity of contemporary (consumerist) culture. Everything must work or operate for the benefit of humanity. One may say that the possession of private jets, the building of mega-churches covering kilometers of land, and the like only contribute to the ongoing ecological crisis.

Although God is all-powerful, God has also placed human salvation in human hands. In responding to the problem of suffering and, by extension, to ecological woes, Pentecostalism offers an alternate response by shifting the focus to the exercise of human freedom. In Pentecostalism, the exercise of human freedom, especially in the context of prayer and spiritual warfare, leads to taking authority in the name of the Lord and having dominion over sickness, poverty, and every form of suffering. As a result of the believers' union with Christ, they see themselves as ones who overcome.

In the following section, I will summarize how Africans understood the creation prior to the advent of Christianity. After this summary, I will mention the problems arising from the foregoing notions of divine providence.

■ An African Ritualistic View of Creation and Divine Intervention

> The village in the East of Ghana where I grew up was located close to a forest and a river. In the forest from ancient times onwards the ancestors live, therefore it was sacred. In the river there lived the spirit of the water, therefore it was sacred as well. Then people of my village became Christians. Now, according to the new Christian worldview, there were no ancestors any more in the forest, and also there were no spirits any more in the river. The taboos were disintegrating and disappearing. Instead the people started to make use and exploit both the forest and the water of the river for their own purposes. Today, next to this village there is no forest left anymore and the river—it turned into a cesspool. Who has made a significant mistake here? And for what reason?[30]

Africans have often been described as "notoriously religious." This may be because an African cosmology is typically ritualistic. Whether this is the case today is a different matter. The entirety of life in pre-Christian Africa was lived and shaped by one's belief in a spiritual worldview. Although this has probably changed, the majority of Africans still believe that life implies spiritual warfare. Whatever happens must have happened spiritually before happening physically. When a predicament befalls someone, the question is not what happens but who did it. Not only this, but nature is also considered sacred. As a result, many Africans had a

---

30. See Werner, "The Challenge of Environmental and Climate Justice," 52, quoting an orally transmitted story from Emmanuel Anim of the Church of Pentecost, Ghana.

tripartite worldview: the earth, the heaven above, and the earth beneath. Everything that exists below the earth, on the earth, and above the earth was sacred.[31] Although not pantheistic, this view held God's immanence in the cosmos in high regard.[32]

Growing up as a village boy and traveling along a rocky footpath amid a fearful forest was both dreadful and puzzling. On the one hand, the sight of the rocks and magnificent forest was a source of wonder. On the other hand, it was a daily nightmare. This is because such vegetation poses two challenges: firstly, there is its sacredness and the need for it to be hallowed and secondly, it was also believed that spirits inhabit some of the vegetation, especially mahogany trees. When coming back from the school or market and even the farm, the sight of a mahogany tree was dreadful. Given such a ritualistic notion of the creation, humans had the moral obligation to preserve creation.[33] In fact, it was considered a taboo and grievous offence to cut down any of those trees or to fish in a sacred river.[34]

African Traditional Religions (ATRs), especially in West Africa, would subscribe to the notion that creation is God's temple. Mountains, rivers, rocks, trees, hills, and lakes, together with some animals, were considered sacred. These creatures, especially forest and water bodies, were understood as temples. As they were sacred, desecrating them was always accompanied by punishment. Understandably, no religious worshipper would want the temple of their deity to be desecrated.[35] Many tribes in Nigeria, for instance, had penalties for open defecation because it was taboo to desecrate Mother Earth. Hunting and fishing were regulated.[36]

Further, bush burning was forbidden because it could result in some animal and plant species going into extinction. Additionally, because people in those days harvested their farm produce and kept them on the farm, bushfires could result in severe loss. After all, they believed that the best way to maintain soil fertility was through the avoidance of bush burning and by shifting cultivation from one portion of land to another.

---

31. Olusakin and Udoh, "African and Christian Theology of Environment," 76.

32. One of the creation myths in central Nigeria among the Tiv people of Benue State depicts God (Aondo) as always present within creation. This myth suggests that heaven and earth were not separated. The distance between heaven and earth was so short that tall persons could touch the sky with their hands. However, this close proximity was destroyed when a woman became a nuisance to God when she pounded yam in the night, and the piston continually touched the sky. God was aggrieved by such disrespectful behaviour and expanded the distance between heaven and earth.

33. Dick, "Religion and the Control of Environmental Crises in Nigeria," 306.

34. See Awajiusuk, "Indigenous African Environmental Ethics."

35. Chinedu, "Religion and the Control of Environmental Crises in Nigeria," 307.

36. Oulusakin and Udoh, "African and Christian Theology of Environment," 76–77.

This ritualistic view of creation was connected to the belief in pre-Christian African religion in a transcendent Creator. The Creator did not just create and abandon the universe as in some forms of deism. On the contrary, everything is under God's control.[37] The belief in divine providence can be extrapolated from African folklore, stories, and even names. Names such as Azhiwointa (God heard us), Azhimuhi (only God knows all), Azhinawre (God always does well), Asidaiazhi (it is God's saying), Azhisei (there is God), and Azhininta (God gave us) (among the Ninzo tribe of North Central Nigeria) signify the ever-present divine finger in human affairs. Although this is the case, the belief in the existence of evil forces meant that some predicaments were not regarded as punishment from God but as the work of the enemy.

The foregoing may explain why there were fewer ecological problems before the advent of Christianity. Although the likes of Harvey Sindima[38] have argued in line with Lynn White[39] that Christianity, with its anthropocentric view of the creation, is the primary catalyst of the ecological crisis today in Africa, I doubt that this is the only reason. We cannot deny that the way Christians appropriated certain theological doctrines affected a sustainable way of living.

Let us now examine how one's view of the divine providence could impact one's responses to the ecological crisis.

∎ Problems Arising from Current Appropriations of Divine Providence

In this section, I will address a number of problems arising from notions of divine providence. I will show how an inadequate appropriation of the classical doctrine of providence (theocentrism) has led to a reluctance to curb the climate crisis, also in Nigeria. I will also show how other Christians take a stand that might be in line with the Wesleyan/open-theistic notion of divine providence (domineering appropriations) that allows humans some freedom to do with creation as they would want.

Michael York believes that the catalyst of the "Anthropocene" period is religion. He argues that the "Anthropocene discourse [...] taps into millenarian hopes and fears, and [it] draws from ancient reservoirs of prophecy, theodicy, and eschatology."[40] In other words, the Abrahamic

---

37. See Kanu and Ndubisi, "On Divine Providence in African Cosmology," 69.

38. See Sindima, "Community of Life."

39. See White, "The Historical Roots of Our Ecological Crisis."

40. York, "Religion and the Environmental Crisis," 13.

religions provide the breeding ground for ecological woes, particularly in their discussion on millennialism. These Abrahamic religions look back to prophecies about the supposed end of the world and the question of God's justice in the problem of evil. What follows from this is the issue of how to understand the ecological crisis from an eschatological point of view. Further, another issue is understanding climate degradation from the perspective of the theodicy problem.

Further, it is argued that the Abrahamic religions hold conflicting ethical views about creation. On the one hand, they seem to speak about the dignity of creation. However, on the other hand, they are unable to uphold the dignity of creation. In other words, this raises ethical questions and objections. "[... P]ollution and destruction of [the] environment are religious and ethical problems that derive basically from irreverent and immoral attitudes toward nature, rather than from technological inadequacy alone."[41] In what follows, I will discuss theocentrism, eschatology, and domineering views of the human-creation relationships and their implications for the ecological crisis, as consequences of the doctrine of divine providence in ecological crisis.

Theocentrism

Many contemporary thinkers believe that religion[42] in general and monotheism in particular[43] is the root cause of the ecological crisis, while others limit this accusation to classical theism. For example, some ecofeminist theologians believe that Christian monotheism has provided a climate that allowed for ecological destruction:

> According to ecofeminist theologians, the monotheist conception, which is claimed to represent the dominant understanding of God, especially in Western Christianity, supports a hierarchical God and universe relationship as a result of the application of the dualist perspective in explaining reality. In this hierarchical structure, God is portrayed as a being outside and beyond this world, who legitimizes power relations and enables the weak to be oppressed.[44]

---

41. Barbour, *Earth Might Be Fair*, 10, quoted in York, "Religion and Environmental Crisis," 13.

42. See White, "The Historical Roots of Our Ecological Crisis."

43. Hava Tirosh-Samuelson argues that "Precisely because nature is created, Judaism does not take nature to be inherently sacred or worthy of veneration. Instead, nature is viewed as imperfect, requiring human management and care: only human actions in accord with divine commands sanctify nature, making it holy." When taken seriously at face value, this conclusion gives a foundation to the domineering model of the human-nature relationship, which seems to aggravate the ecological crisis. See Tirosh-Samuelson, "Judaism," 1.

44. Quoted in Öztürk, "Is Monotheism the Root Cause of the Ecological Crisis?" 302.

The doctrine of divine providence is the melting point where this accusation is being distilled. Given that classical theists situate the doctrine of divine providence in their understanding of the doctrine of creation, it has been criticized for being anthropocentric, hierarchical, and therefore the root of ecological crisis. Lynn White, in his famous thesis on the historical roots of ecological crisis, blames the roots of the ecological crisis on the command in Genesis 1, where God gave humanity dominion (*radah*) over nature. White argues that a few elements in medieval Christian theology and religious values encouraged humans to assert such dominion over nature aggressively.[45]

The classical doctrine of providence, especially in Augustine and Calvin, seems to leave no room for human freedom. Classical theists do speak about the compatibility of divine action and human freedom. However, human freedom can accomplish little when everything is overruled by the divine power. Meticulous divine providence seems to offload the burden of care for the creation from humanity's shoulder and places it solely on God's. Such a theocentric view may lead to a lack of moral accountability and therefore to more suffering, also among nonhuman animals.

With the rise of African Pentecostalism and AICs, the dominance of Western Christianity (and probably classical theism) has been challenged. Most of the mega-churches in southern Nigeria are Pentecostal in orientation. The presence of such churches may be found throughout the African continent, Europe, and America.

Because everything in the traditional African worldview is not only sacred but viewed through a spiritual lens, every predicament, whether cosmological/natural or moral, is believed to have a spiritual cause. The presence of natural disasters and every form of misfortune are part of the fabric of the society. Every person in society must adapt to this reality so as to overcome these disasters or misfortunes. Given such a pre-Christian notion of reality, many Africans came to the faith but were unable to eradicate such beliefs. In this context, the climate crisis is seen as a form of punishment from God. It is normal for Christians in Nigeria to fast when there is no rain or when there is an outbreak of locusts devouring their crops.[46]

In a study by George Christian Nche, he presented the following five theses:

1. God created this world, and it belongs to God.
2. If we care for life, we must care for the environment.

---

45. White, "The Historical Roots of our Ecological Crisis," 7.

46. See Izidor and Igwe, "Climate Change and the Church," 378.

3. Our humanity depends upon the environment.
4. Creation is bound up with salvation.
5. The earth crisis is a crisis of culture to which the gospel speaks.[47]

Given the foregoing, one may expect a lot from Pentecostal churches in Africa. Unfortunately, most of the presiding pastors of such churches say little or nothing about ecological concerns. This leads Nche to conclude that his study "[...] highlight[s] the culpability and indifference of churches towards the challenges of climate change."[48]

One of the respondents interviewed by Nche says that "[i]f I could say, in my own opinion, [...] we can't attribute the (climate) change to a particular individual or a particular thing [...] it was meant to be so, because even if you go scripturally, they say, there is time for everything. Therefore, there is always cause for change."[49] In other words, although humans have been contributing to ongoing ecological destruction, it has nothing to do with human action.[50]

Eschatological Appropriations

A second reason for the alleged complicity of Christianity in ecological destruction, which follows from the first and seems to receive backing from both Pentecostals and African evangelical churches, is the notion of eschatology. Most Christians in Africa, and especially Nigeria, seem to believe that the parousia is around the corner. Many Christians in Africa believe that we are living in the end times and that ecological degradation is a testament to that. In dealing with the current climate crisis, both Pentecostal and mission-founded churches in Nigeria seem to embrace such a view of eschatology. They believe that because the creation groans and is awaiting its redemption, it must be that the current crisis is the process through which this redemption will be realized. Therefore, there is nothing we can do about it. After all, one of the signs of the parousia, according to such Christians, is that God will unmake the creation. Ecological crises, including wars and famine, are pointers that the end is near. In a paper published recently, four authors, claiming ample biblical support, argued that "[...] climate change from a Christian perspective is the sign of the end of the world and the second coming of Jesus Christ (parousia)."[51]

---

47. Nche et al., "Challenges of Climate Change and the Culpability of Churches," 177.

48. Nche et al., "Challenges of Climate Change and the Culpability of Churches," 170.

49. Nche, "Beyond Spiritual Focus," 155.

50. Nche, "Beyond Spiritual Focus," 154.

51. Agujiobi et al., "Climate Change as an Act of 'God' or 'Man'," 1.

These authors continue to argue that their paper aims to "[...] prove that both the physical and social economic effects of climate change on the Earth's surface have some spiritual undertones, which are interpreted by Christians to mean the signs of the end of the world."[52] In other words, "the overall aim of this paper is to show that climate change and its effects on the Earth's surface are ordained by God and are signs of the end time."[53]

These authors cite Matthew 24:3-7, 1 Corinthians 15:51, 2 Thessalonians 4:13-14, 2 Peter 3:3-12, and Mark13:3-8 in support of their arguments. They argue that other signs of the parousia include ecological disasters, famine, the outbreak of diseases, conflict, and wars, primarily over limited resources arising from the ecological crisis.[54]

Samuel Izidor and Andrew Igwe follow a similar path. They add that "[o]bservedly, the sin agency theory [sic] continues to appear with its accompanying punishment as shown by the expositions on the various destructions executed by God upon man and the earth, portending environmental and climatic changes with serious implications for man."[55] Nevertheless, they maintain that humans, notwithstanding the foregoing, could labor to curb the ecological menace.[56]

It is evident that many Pentecostal pastors are ignorant or uninterested in the ecological woes associated with the "Anthropocene." This is in line with a study by Pepper and Powell (2013). They reveal that "Pentecostal churches and their clergy typically occupy last or close to last position on a range of measures of environmental activity [...]."[57] In a study conducted by Nche, it was established that many Pentecostal Christians and pastors are unaware of the causes of the climate crisis.[58] This is not just a matter of lack of knowledge but probably a matter of priority. The ecological crisis, just like the problem of evil, is nonexistent to the average Pentecostal Christian, given that in Christ, those things do not exist except for those who lack faith. In the mission-founded churches, the same problem prevails. I do not remember taking a course which highlighted the ecological crisis during my days in the seminary.

---

52. Agujiobi et al., "Climate Change as an Act of 'God' or 'Man'," 10.

53. Agujiobi et al., "Climate Change as an Act of 'God' or 'Man'," 10.

54. Agujiobi et al., "Climate Change as an Act of 'God' or 'Man'," 10-11.

55. Izidor and Igwe, "Climate Change and the Church," 383.

56. Izidor and Igwe "Climate Change and the Church," 378.

57. Nche, "Beyond Spiritual Forces," 155.

58. See Nche, "Beyond Spiritual Forces," 155.

This may explain why, in the study conducted by Nche, one of the respondents argued that the "[...] church should focus on salvation rather than climate change awareness [...] two others stated that the 'awareness' the church should create is to let people know that climate change is a sign that the end is near."[59] This conclusion that our focus should be on the salvation of lost souls rather than on the ecological crisis raises the question of how salvation is then understood. This may well affect an environmental concern negatively.[60]

Domineering Appropriations

In contrast to the ritualistic notion of creation upheld by worshippers in ATRs, where creation is considered sacred, Pentecostal churches call for the sanctification of ecology by casting out marine spirits from streams and rivers and cutting down evil forests. In ATRs, some individuals are able to manipulate the spirits either positively or negatively. This finds resonance in Pentecostalism because of the emphasis on human freedom. A notable church founder and preacher from Nigeria, Pastor Chris Oyakhilome, believes that, given Christians' new position in Christ, they can alter divine decrees.[61] Many Pentecostal members and preachers also hold this radical idea. They believe that God is all-powerful and has granted them the right to take authority in the name of Jesus to command whatever they desire to happen to the extent of requesting those things that are contrary to divine decrees. As God is bound by God's covenant, God cannot but grant them their requests. After all, "what God cannot do does not exist." Such an ability to change divine decrees has implications for ecological woes. One would expect them to leverage their emphasis on freedom to formulate a viable response to the climate crisis. However, this seems not to be the case.

As stated in the introduction, not only Pentecostal pastors call for the sanctification of the environment. Countless stories may be found where pastors ordered the destruction of nature in the name of cleansing. The case of my local village chief is an example of this domineering/sanctifying approach to the natural environment. The fundamental force behind such actions is the belief that humanity, together with God, can determine what happens to creation. God is not the sole determinant. God has made us little gods who can determine our fate and that of the creation.

---

59. Nche, "Beyond Spiritual Forces,"157.

60. See Nche, "The Church Climate Action," 8.

61. See Oyakhilome, *Your Rights in Christ*, 15–18.

◼ Divine Providence and the Ecological Crisis: Some Prospects

Both the classical and the open theistic notions of divine providence present challenges in dealing with the current ecological crisis. It seems that it is not only the type of divine providence which one holds that determines one's response to the crisis but also how one views creation.

Ordinarily, classical theism is more comforting to me as an individual while considering viable responses to the climate crisis.

First, the idea that creation is a product of divine grace necessitated by God's love and God's desire for fellowship with the creation connotes that there is a divine plan for creation. If that is the case, then we cannot expect God to govern it minimally. In fact, whether one holds that God exercises meticulous or general providence, as in the case of open theism, in the end, one must accept that natural laws are put in place by God. Therefore, God is still controlling everything.

Second, the idea that exhaustive divine providence undermines human freedom is debatable. Augustine (as shown above), for instance, argues that God's knowledge of an action does not confer necessity or limitation on the said action but precisely allows such actions to be free. As far as Calvin is concerned, divine providence does not excuse human ignorance. It does not encourage a lack of prudence and therefore does not call us to mistreat creation, because it is the theater of God's glory. The honor accorded to creation by Calvin does not seem to encourage the abuse of creation but rather a call to preserve it.[62]

Third, although exhaustive divine providence has been dismissed as problematic for infringing on human freedom, this seems not to be the case, if noting that Calvin's doctrine of election and predestination formed the basis for Max Weber's concept of capitalism. If the doctrine of election, which seems more fatalistic than divine providence, could challenge believers to be prudent in managing their resources, divine providence, which gives room for human responsibility, could do more in curbing the climate crisis. After all, the classical doctrine of divine sovereignty has "[…] never been intended to take human responsibility lightly; it rather served to ground and undergird such responsibility. If indeed we are to submit to God's sovereign will, this means that all our activities should be directed to what we can know about God's purposes."[63] And this includes taking our responsibility to care for the creation seriously. Here, the benefit of freedom

---

62. Calvin, *Institutes*, 1.17.5.

63. Van den Brink and Van Urk, "Climate Change and God's Work of Election," 458.

as upheld in open theism, and by extension, accountability, might be helpful in combating the ecological crisis.

Fourth, contrary to open theism, whether God's work of creation was *ex nihilo* or not, creation has a *telos* in the divine plan. And just as in election and predestination, believers are encouraged to seek the lost ones, so in a teleological understanding of creation, Christians are encouraged to seek the salvation of creation, given that, according to the Bible, creation got itself into a mess as a result of human rebellion against God. So if God is redeeming humanity and guiding us in our earthly journey towards the parousia, God expects us to also work for the salvation of creation, which might be fully achieved only in the eschaton.

Fifth, the classical doctrine of providence gives hope and comfort amid the climate crisis. Amid our struggles to heal the Earth of pollution, we can feel the comforting presence of God and remember the assurance that God is with us always.

Sixth, although we may be at the end of the age and moving towards the parousia, we must also do everything within our reach to ensure the salvation of creation. In the end, just as the moral problem of evil does not disprove God's love, power, and divine control, the same applies to ecological woes in the "Anthropocene."

Finally, divine immanence, as conceived in the African ritualistic view of the cosmos is more promising in combating ecological destruction than a view of a transcendent, domineering creator who either single-handedly overrules everything for their glory or the God who grants a high degree of freedom to humans to the extent that we can both change divine decrees and do with creatures as we wish. Both extreme notions either lead to freezing out or an undue dominance over the world. The African notion that creation is sacred and should be respected resonates perfectly with the current view of the cosmos that calls humanity to care for, preserve, and serve the creation.

■ Bibliography

Agujiobi, Nnaemeka Emmanuel, et al. "Climate Change as an Act of 'God' or 'Man': An Eschatological Account." *Ikenga: Journal of African Studies* 23:2 (2022) 1–17. https://doi.org/10.53836/ijia/2022/23/2/009

Antombikums, S. Aku. "African Pentecostalism and Its Dilemma." In *African Pentecostalism*, edited by Lekan Diaro, Margaret Ojo and Tukumbo Bankole, 4–16. Ede: Redeemer University Press, 2022.

———. *Divine Control, Human Agency and the Problem of Evil* (May 2025 forthcoming).

———. "Open Theism and the Problem of Evil." PhD thesis, Vrije Universiteit Amsterdam, 2022. https://research.vu.nl/en/publications/open-theism-and-the-problem-of-evil.

Arminius, Jacob. *The Works of James Arminius*. 3 Vol. Grand Rapids: Christian Classics Ethereal Library, 2002.

Augustine, St., *The Retractations*. Translated by Mary Inez Bogan. Downers Grove: The Catholic University of America Press, 2017.

Awajiusuk, Finomo Julia. "Indigenous African Environmental Ethics: A Panacea for Sustaining the Niger Delta Environment." *Journal of Religion and Culture* 10:1&2 (2010) 101–20.

Barbour, G. Ian, ed. *Earth Might Be Fair: Reflections on Ethics, Religion, and Ecology*. Hoboken: Prentice Hall, 1972.

Calvin, John. *A Defence of the Secret Providence of God: By Which He Executes his Eternal Decrees: Being A Reply To the "Slanderous Reports" (Rom. 3:8) of a Certain Worthless Calumniator Directed against The Secret Providence of God*. Translated by Henry Cole. Waterford: CrossReach, 2018.

———. *Institutes of the Christian Religion*. Edited by John T. McNeill, translated and indexed by Ford Lewis Battles. Louisville: Westminster John Knox, 1960.

Conradie, Ernst M. "The Story of God's Work: An Open-Ended Narrative." In *T&T Clark Handbook of Christian Theology and Climate Change*, edited by Ernst M. Conradie and Hilda P. Koster, 462–66. London: T&T Clark, 2020.

Crisp, Oliver D., and Fred Sanders, eds. *Divine Action and Providence: Explorations in Constructive Dogmatics*. Grand Rapids: Zondervan Academic, 2019.

Dick, Chinedu Jonathan. "Religion and the Control of Environmental Crises in Nigeria." In *African Eco-Theology: Meaning, Forms and Expressions*, edited by Ikechukwu Anthony Kanu, 303–19. Milton Keynes: AuthorHouse, 2022.

Elliot, Mark. *Providence Perceived: Divine Action from a Human Point of View*. Boston: Walter de Gruyter, 2015.

Fergusson, David. *The Providence of God: A Polyphonic Approach*. Cambridge: Cambridge University Press, 2018.

Guerlac, Henry, and M. C. Jacob. "Bentley, Newton, and Providence: The Boyle Lectures Once More." *Journal of the History of Ideas* 30:3 (1969) 307–18. https://doi.org/10.2307/2708559

Helm, Paul. *The Providence of God: Contours of Christian Theology*. Downers Grove: Intervarsity Press, 1994.

Hume, David. *An Enquiry Concerning Human Understanding*. Oxford: Oxford University Press, 2000.

Izidor, Samuel, and Andrew A. Igw. "Climate Change and the Church: An Eschatological Perspective." *Randwick International of Social Sciences Journal* 3:2 (2022) 377–87. https://doi.org/10.47175/rissj.v3i2.347

Kanu, Ikechukwu Anthony, and Ejikemeuwa J. O. Ndubisi. "On Divine Providence in African Cosmology." In *African Philosophy: Contemporary Issues and Perspectives*, edited by Anthony Kanu Ikechukwu and Ejikemeuwa J. O. Ndubisi, 57–72. Silver Springs: The Association for the Promotion of African Studies, 2021.

Lloyd, Genevieve. *Providence Lost*. Cambridge: Harvard University Press, 2008.

Mann, William E. "Augustine on Evil and Original Sin." In *The Cambridge Companion to Augustine*, edited by Eleanore Stump and Norman Kretzmann, 40–48. Cambridge: Cambridge University Press, 2006.

Muller, Richard. *Divine Will and Human Choice: Freedom, Contingency, and Necessity in Early Modern Reform Thought*. Grand Rapids: Baker Academic, 2017.

Nche, George C. "Beyond Spiritual Focus." *Weather, Climate, and Society* 12:1 (2020) 149–69. https://doi.org/10.1175/WCAS-D-19-0001.1

———. "The Church Climate Action: Identifying the Barriers and the Bridges." *Transformation: An International Journal of Holistic Mission Studies* 37:3 (2020) 1–20. https://doi.org/10.1177/0265378820931890

Nche, George C., et al. "Challenges of Climate Change and the Culpability of Churches: Towards an Effective Church Climate Change Action in Nigeria." *Missionalia* 45:2 (2017) 168-87. https://doi.org/10.7832/45-2-183

Olupona, Jacob K. "West Africa." In *The New International Dictionary of Pentecostal and Charismatic Movements*, edited by Samuel M. Burgess. Revised and expanded edition, 61-82. Grand Rapids: Zondervan, 2008.

Olusakin, Vincent A., and Moses Udoh. "African and Christian Theology of Environment as a Model for the Control of Global Warming." *Ifiok: Journal of Interdisciplinary Studies* 4 (2018) 70-86.

Oord, Thomas. "God's Initial and Ongoing Creating." In *T&T Clark Handbook of Christian Theology and Climate Change*, edited by Ernst M. Conradie and Hilda P. Koster, 362-72. London: T&T Clark, 2020.

Oyakhilome, Chris. *Your Rights in Christ*. Ikeja: Loveworld, 1998.

Öztürk, Sevcan. "Is Monotheism the Root Cause of the Ecological Crisis? Ecofeminist Conceptions of the God-Universe Relationship." *Kader* 21:1 (2023) 301-19. https://doi.org/10.18317/kaderdergi.1255426

Pinnock, Clark, et al. *The Openness of God: A Biblical Challenge to the Traditional Understanding of God*. Downers Grove: Intervarsity, 1994.

Pepper, Miriam and Powell, Ruth. "Senior Local Church Leaders' Environmental Views and Actions." *NCLS Research (Occasional Paper 21)*, Mirrabooka Press, Adelaide, (2013), 1-7.

Rice, Richard. *God's Foreknowledge and Man's Free Will*. Minneapolis: Bethany, 1985.

Sanders, John. *The God Who Risks: A Theology of Divine Providence*. Downers Grove: Intervarsity, 2007.

———. "Open Creation and the Redemption of the Environment." *Wesleyan Theological Journal* 46:4 (2010) 141-49.

Santmire, H. Paul. *The Travail of Nature: The Ambiguous Ecological Promise of Christian Theology*. Minneapolis: Fortress, 1985.

Sindima, Harvey. "Community of Life: Ecological Theology in African Perspective." In *Liberating Life: Contemporary Approaches in Ecological Theology*, edited by Charles Birch, William Eaken and Jay B. McDaniel, 137-47. Maryknoll: Orbis, 1990.

Tirosh-Samuelson, Hava. "Judaism." In *Oxford Handbook of Religion and Ecology*, edited by Roger S. Gottlieb, 1-39. Oxford: Oxford University Press, 2006.

Van den Brink, Gijsbert, and Eva van Urk. "Climate Change and God's Work of Election." In *T&T Clark Handbook of Christian Theology and Climate Change*, edited by Ernst M. Conradie and Hilda P. Koster, 451-61. London: T&T Clark, 2020.

Wariboko, Nimi. "Pentecostalism in Africa." In *Oxford Research Encyclopedias*, edited by Thomas Spear, 1-25. Oxford: Oxford University Press, 2017. https://doi.org/10.1093/acrefore/9780190277734.013.120

Werner, Dietrich. "The Challenge of Environment and Climate Justice: Imperatives of an Eco-Theological Reformation of Christianity in African Contexts." In *African Initiated Christianity and the Decolonisation of Development: Sustainable Development in Pentecostal and Independent Churches*, edited by Philipp Öhlmann, Wilhelm Gräb, and Marie-Luise Frost, 51-72. London: Routledge, 2020.

White, Lynn, Jr. "The Historical Roots of our Ecological Crisis." *Science* 155 (1967) 1203-7. https://doi.org/10.1126/science.155.3767.1203

York, Michael. "Religion and the Environmental Crisis." In *Ecotheology Sustainability and Religions of the World*, edited by Levente Hufnagel, 7-28. London: IntechOpen, 2023. https://doi:10.5772/intechopen.104002

Common Grace and Sustainability: Some South African Reformed Perspectives

Ernst M. Conradie[1]

◼ Telling the Story Still Hinges on the Noah Story

"How could the suffering of God's creatures in the 'Anthropocene' be reconciled with trust in God's loving care?" In the Reformed tradition, such a question would typically be addressed in terms of reflection on God's providence. Accordingly, it is precisely through God's providence that God is making room for the story to continue. This story hinges on Genesis 6:5-8:

---

1. Ernst M Conradie is senior professor in the Department of Religion and Theology at the University of the Western Cape.

How to cite: Conradie, EM 2024, 'Common Grace and Sustainability: Some South African Reformed Perspectives', in EM Conradie & UL Vaai (eds.), *Making Room for the Story to Continue?*, An Earthed Faith: Telling the Story amid the "Anthropocene", vol. 4, AOSIS Books, Cape Town, pp. 55-88. https://doi.org/10.4102/aosis.2024.BK415.03

⁵ The Lord saw that the wickedness of humans was great in the earth and that every inclination of the thoughts of their hearts was only evil continually. ⁶ And the Lord was sorry that he had made humans on the earth, and it grieved him to his heart. ⁷ So the Lord said, "I will blot out from the earth the humans I have created—people together with animals and creeping things and birds of the air—for I am sorry that I have made them." ⁸ But Noah found favor in the sight of the Lord [...] (New Revised Standard Version, Updated Edition [NRSVUE]).

In short, following the spread of evil in the world, God decided (as per Gen 6:7) to destroy[2] humans and with them all other living creatures. Although Noah's flood is portrayed as a primeval happening and clearly not as a historical event, it is well situated amid the melting ice that flooded Mesopotamia at the advent of the Holocene. In the aftermath of the destruction of the flood, as the story goes, God started anew through a covenant with Noah, his family, and indeed all living creatures who exited from the ark (Gen 9:10) that would allow the descendants of Noah to "be fruitful and multiply" and to "abound on the earth" (Gen 9:7).

As the subsequent chapters of Genesis indicate, God's covenant with Noah did not quite go according to plan. Following the rise of imperialism symbolized by Babel (Gen 11), God made yet another plan, namely through the election of Abram. Then follows the history of salvation from Abraham to Moses and the exodus from Egypt, from David to the Babylonian exile, from the failures of reconstruction and of becoming a light for all nations to the coming of the Messiah, and from the formation of the church to God's mission to reach the ends of the earth. In hindsight we know that this story of Israel, of the church, and of God's mission did not quite go according to plan either (to put this mildly), because evil continued to spread despite (and often because of) the advent of Christianity. The impact of such evil is now evident in the so-called "Anthropocene" and will be embedded in the Earth's rock layers for millions of years to come.

Arguably, the story of Noah continues to form the backdrop of the history of salvation. The ark itself may be a symbol of rescue and therefore redemption, but by itself this does not provide a long-term solution. This is why the covenant with Noah is crucial. Without the covenant with Noah, the rest of the story would not be possible. Clearly, this covenant is not to be equated with salvation, but it made room for the story to continue. Despite the change in food regulations (Gen 9:1–6) it provided parameters to minimize violence so that all living creatures could flourish (the דָּם, i.e., "blood" in Gen 9:4 being the symbol of life). It also serves as God's way of self-restraining so that God would not again destroy the world that

---

2. The Hebrew word אָמְחָה in Genesis 6:7 can be translated as destroy or blot out, as in blot out from God's memory (see Ex 17:14). Karl Barth understands this as abandoning creation to chaos, to annihilation or self-annihilation. See his *Church Dogmatics III.3*, 78, and the critique of Barth's position by Van Ruler below.

God created. It puts a self-check on God.³ There is also the enigmatic figure of the "restrainer" (κατέχων, often understood as the state) who holds back the spread of evil (2 Thes 2:7) in anticipation of Christ's παρουσία.⁴

To add "arguably" here is to acknowledge that not all Christians follow this story-line. In Christian ecotheology, several different story-lines are adopted. Much of this hinges on the question whether the main problem is regarded as the spread of human sin resulting in structural violence, or whether natural suffering is emphasized, or the interplay between them. In short, is the problem sin as guilt, sin as power, or the tragic dimension of creaturely existence? What is it that we need to be saved from? These are contrasting visions,⁵ where deep confessional divides on nature and grace are still at play.⁶ I will return to this, but I need to acknowledge my Reformed roots in this regard.⁷

A formative expression of a Reformed confession of God's providence is found in Article 13 of the Belgic Confession (1561). It is worth quoting here in full:

> We believe that this good God, after creating all things, did not abandon them to chance or fortune but leads and governs them according to his holy will, in such a way that nothing happens in this world without God's orderly arrangement. Yet God is not the author of, and cannot be charged with, the sin that occurs. For God's power and goodness are so great and incomprehensible that God arranges and does his works very well and justly even when the devils and the wicked act unjustly. We do not wish to inquire with undue curiosity into what God does that surpasses human understanding and is beyond our ability to comprehend. But in all humility and reverence we adore the just judgments of God, which are hidden from us, being content to be Christ's disciples, so as to learn only what God shows us in the Word, without going beyond those limits. This doctrine gives us unspeakable comfort since it teaches us that nothing can happen to us by chance but only by the arrangement of our gracious heavenly Father, who watches over us with fatherly care, sustaining all creatures under his

---

3. See Havea, "Covenant," 54.

4. For a discussion, see Moltmann, *Ethics of Hope*, 13–15.

5. Arnold van Ruler maintains that these visions are contradictory: "Here there is a contradiction between the Greek (and generally pagan) and the Christian analysis of being human. The one excludes the other: tragedy and guilt." See *This Earthly Life Matters*, 168.

6. Compare, for example, discourse on common grace with the positions adopted by Matthew Fox and Joseph Sittler. In his *Essays on Nature and Grace*, Joseph Sittler explores the significance of (saving) grace for transforming nature, but does not employ the notion of common grace. By contrast, in *Original Blessing*, Matthew Fox seeks to understand God's work of creation as the primary form of grace so that creation and salvation tend to become fused. One may say that there is simply no ecumenical consensus on the relation between nature and grace, also not in the field of Christian ecotheology.

7. I have discussed these contrasting visions at some length in a project on "Redeeming Sin?" See the monograph with that title and my reflections on the project in "The Project and Prospects of 'Redeeming Sin?'"

lordship, so that not one of the hairs on our heads (for they are all numbered) nor even a little bird can fall to the ground without the will of our Father. In this thought we rest, knowing that God holds in check the devils and all our enemies, who cannot hurt us without divine permission and will. For that reason we reject the damnable error of the Epicureans, who say that God does not get involved in anything and leaves everything to chance.[8]

In the subsequent Reformed tradition, such provision, re-described as God's providence, is traditionally unpacked with reference to the three dimensions of God's work of conservation (*conservatio*), God's governance in history (*gubernatio*), and the synergy between divine, human and other forms of agency (*concursus*).[9]

In this essay, I will focus on God's work of conservation with the hope of addressing the other two themes elsewhere.[10] More specifically, I will explore the notion of "common grace" that is employed in one branch of the Reformed tradition. The question that I will pose here is whether the notion of common grace needs to be discarded together with any reference to the orders of creation, or whether this remains a necessary precondition for any retrieval of the notion of God's providence. Moreover, can common grace be employed to capture the theological significance of contemporary secular discourse on sustainability, or should it be completely discarded? I will even entertain the question whether discourse on sustainability itself may need to be discarded given the shift from the Holocene to the "Anthropocene." If Holocene stability is no more, what then? Is the danger not that discourse on sustainable development seeks to sustain an unjust global economic order as long as possible? Is it not necessary to disrupt such order? Is discourse on ecojustice (from the Global South) compatible with discourse on sustainability (from the Global North), or do these represent contrasting visions? Put bluntly: is God always creating order from chaos, or is God perhaps causing chaos to allow for a new dispensation to emerge "from the edges of chaos"?[11]

---

8. For this translation of the Belgic Confession, see https://www.crcna.org/sites/default/files/BelgicConfession.pdf [last accessed October 10, 2024].

9. For a thorough recent discussion of the doctrine of providence in the Reformed tradition, see Fergusson, *The Providence of God*, 77–109. My contribution here is constructive more than historical.

10. Essays on *gubernatio* and *concursus* are included (together with this essay) in a companion volume, entitled *A God Who Cares? Reformed Perspectives on God's Providence amid the Shift from the Holocene to the "Anthropocene,"* accepted for publication by AOSIS.

11. The allusion is to chaos theory (employed in thermodynamics but also in evolutionary biology) that allows for the possibility of new forms of order to emerge (through self-organization) "from the edge of chaos," i.e., in states far from equilibrium. I will not explore the fertile science-and-theology debates on chaos and complexity here.

■ Neo-Calvinism and its Legacy

The notion of common grace is associated especially with the names of Abraham Kuyper (1837–1920) and Herman Bavinck (1854–1921). This Dutch school of "neo-Calvinism" may be understood as an attempt to break out of the rigid scholasticism of Reformed orthodoxy by returning afresh to the theology of John Calvin. If theology is a matter of faith seeking understanding, then faith cannot be fixated on its cognitive content, or on propositional truth claims, but has to be rooted in a relationship based on trust in God's sustained loyalty not only to God's covenant partner but to the whole of creation despite the continuing impact of human sin. Faith is a matter of personal knowledge emerging from an ongoing, dynamic relationship built upon reading the Bible and prayer, exegesis, and proclamation. The content of faith therefore cannot be captured in fixed doctrinal truths (as per orthodoxy or the subsequent Princeton fundamentalism) but requires listening to God's Word, always anew, and reflecting on its meaning, for us, today.

Such rootedness of the Christian faith in personal piety did not preclude but precisely enabled the this-worldly orientation of neo-Calvinism (the mystic enabling the prophetic). Kuyper's widely cited comment is that "There is not a square inch in the whole domain of our human existence over which Christ, who is Sovereign over all, does not cry, Mine!"[12] In its origins, neo-Calvinism was an anti-elitist movement, expressing a somewhat populist concern for the "*kleine luyden*" ("little people") in the Netherlands. This still allowed for a wide range of political sentiments, from the far right (including some of Kuyper's followers while he became the Dutch Prime Minister as leader of the Antirevolutionary Party) to the far left (a sustained concern for social justice).[13] Wherever neo-Calvinism reverts to a rigid Reformed orthodoxy, it seems to play in the hands of the political right, not least in South Africa, since such orthodoxy becomes a handy tool to maintain authority and therefore the status quo.

This branch of the Reformed tradition of Dutch origin had a decisive influence on my own theological education. Many of my teachers studied at the Vrije Universiteit in Amsterdam when the influence of Kuyper and Bavinck was still tangible. The role of supervisors such as Gerrit Berkouwer and the missiologist J.H. Bavinck on a generation of South African Reformed theologians such as Jaap Durand, Johan Heyns, Willie Jonker, Bethel Müller, Hennie Rossouw, Flip Theron, Wentzel van Huyssteen, and Conrad Wethmar cannot be underestimated. Although Berkouwer's theology remained

---

12. This famous comment from his inaugural speech at the Vrije Universiteit Amsterdam (1880) was formulated Christologically. See Bratt, *Abraham Kuyper*, 488.

13. This concern for social justice could be symbolized by Nicholas Wolterstorff's *Until Justice and Peace Embrace*.

rooted in that of Bavinck, he also read Karl Barth with considerable appreciation,[14] over and against some of his more rigidly Reformed colleagues such as Valentijn Hepp. This, too, had a significant influence on Reformed theology in South Africa.

Nevertheless, neo-Calvinism had a disastrous impact in the South African context. Most of the apartheid theologians were also attracted to neo-Calvinism, especially to Kuyper and the philosopher Herman Dooyeweerd. Remarkably, apartheid theology understood itself as theologically orthodox (returning to "the old paths"), in reaction to late nineteenth-century theological and political "liberalism," which was also evident in the then Cape Colony. At the very core of apartheid theology was a particular retrieval of the notion of common grace, understood in terms of the so-called orders of creation that restrain the spread of evil in the world so that God's work of salvation (special grace) has the necessary room to proceed. These orders included in the South African context an emphasis on the categories of race, ethnicity, *volk* (people), and culture. This emphasis on the orders of creation was vehemently critiqued by a number of South African Reformed theologians, precisely by drawing on Barth's theology. In addition to the names mentioned above, one may add those of David Bosch (who studied under Oscar Cullmann in Basel) and John de Gruchy (who was deeply influenced by Dietrich Bonhoeffer).

At a personal level, after my basic training, I refrained from engaging with any Dutch theology, did my doctoral research on the Catholic theologian David Tracy, and then started working on ecotheology. It was only when I started with long-term projects on "Hope for the Earth?"[15] and especially on creation and salvation[16] that I rediscovered the work of Herman Bavinck and, especially, Arnold van Ruler.[17]

■ On Common Grace

The concept of common grace is derived from John Calvin, building on insights from Augustine.[18] Even though Calvin did not use the phrase

---

14. See Berkouwer, *The Triumph of Grace*.

15. See Conradie, *Hope for the Earth*.

16. See, especially, Conradie, *Saving the Earth?*

17. See the volume of essays by Van Ruler, *This Earthly Life Matters* that Douglas Lawrie translated and that I edited.

18. See this quotation from Augustine in Schreiner, *The Theater of His Glory*, 12: "The power, omnipotence, and all-sustaining strength of the Creator is the cause of the subsistence of every creature. And if this power were ever to cease ruling created things, their species would at once cease to be and all of nature would collapse [...] For if God were to withdraw his rule from it, the world could not stand, even for the blink of an eye." The reference is to Augustine, *De Genesis ad Literam*, IV.12.32. One may therefore say that belief in God's providence is the inverse of an affirmation of the contingency of creation.

frequently, this was core to his understanding of God's providence. As a humanist scholar, Calvin shared in a positive appraisal of nature that was typically of the Renaissance—as the expression "theater of his glory" suggests. However, Calvin sensed that the cosmic and societal foundations of the late medieval world had crumbled. In the face of the horrors of overwhelming chaos, Calvin, the refugee, could not put his trust in the inherent stability of cosmic or social order but only in the Word of God. After the fall of humanity, any continuing cosmic or social order depends on God's providence, "on the constant sustaining and restraining providence of God which prevents them from falling into chaos."[19] Common grace is the source of all human virtue and accomplishment, even among unbelievers and despite humanity's radical depravity. Such goodness is therefore ascribed not to humans but to the benevolence of God toward sinful humanity.[20] The order does become embedded in the material world itself and allows for its continuity in time, but without God holding the world in God's hands, moment by moment, such order would collapse.[21] This is the ultimate source of the stability, regularity, and continuity in all of creation.[22]

Likewise, ongoing wars, injustices, tyranny, and revolutions provide ample evidence of the moral disorder in society. God nevertheless reins in (Calvin employs the image of a bridle) such disorder, restraining the wicked and the work of the devil by governing all events in order to establish a just social order. God hinders the spread of evil and permits it only in order to

---

19. Schreiner, *The Theater of His Glory*, 3.

20. See Van Leeuwen, "Herman Bavinck's 'Common Grace'," 36.

21. Put graphically, Calvin (following Basil) sensed that the earth is spheric and not flat. But if so, why does the earth (which is heavier than water or air) not fall? Even if it is flat, how does it float on water? What does it rest upon? For Calvin, that could only be because God prevents it from falling: "Nothing in the world is stable except as it is sustained by the hand of God." See Schreiner, *The Theater of His Glory*, 26–27. The quotation is from Calvin's commentary on Psalm 104:5. Protestant astronomers such as Kepler arguably turned their telescopes to the heavens to find God's hand in the movement of the stars. Newton would later explain this in terms of gravity but still saw gravity as God's instrument. The need for a continuous act of God to preserve each creature became increasingly superfluous given the principle of inertia. As Wolfhart Pannenberg observes, the doctrine of providence as *conservatio* never recovered from this blow and is rarely addressed. Note that the need for such preservation is thus separated from the spread of evil. See Pannenberg, "Providence, God, and Eschatology" in *The Historicity of Nature*, 59.

22. See Schreiner, *The Theater of His Glory*, 22. She adds: "To Calvin, the inherent character of creation was not conducive to order; only a great divine power could preserve the grand orderliness we perceive in the universe. The stability of nature depends upon 'the continual rejoicing of God in his works'. If God ceased to rejoice therein, if he ceased to give vigor to the earth or if he looked upon creation with wrath, the sphere of nature would collapse into disorder. Therefore Calvin believed that behind the beauty of nature lay its fragility, dependence, and precarious nature which required the continual preservation of God, for without his providence the stars would collide, the earth would fall down, and water would gush forth and engulf the earth" (28). It is through the "bridle of providence" that God "restrained the wicked and held nature in check" (36).

ultimately overcome evil, as is epitomized by the cross. As Schreiner puts it, God the Father is no "watchtower divinity" who merely "permits" nature or history to run its course but exercises God's sovereign will in all events.[23] In addition to general providence (common grace), for Calvin there is also special providence in that God controls every distinct movement and particular event, even if our experience may point to the contrary (again as epitomized by the cross). If so, God may use our adversity to bring us closer to God.[24]

This notion of common grace was retrieved by Abraham Kuyper, the formidable Dutch pastor, journalist, theologian, and politician.[25] In the massive three volumes of meditations collected in *De Gemeene Gratie* he sketched how such common grace was still at work amid the rapid social change of the late nineteenth century, to restrain evil, to maintain some stability, and to make room for God's special grace (on which he also wrote extensively). He employed the term re-creation (*herschepping*) following God's compassionate decision after the fall of humanity not to destroy creation, neither to replace creation with a completely new creation, but to restrain the spread of demonic powers and catastrophic destruction in a creation still beloved by God.[26]

Such a notion of common grace was further developed by Kuyper's erudite colleague Herman Bavinck, especially in two essays translated as "Common Grace" and "Calvin and Common Grace." Following Calvin, Bavinck describes common grace in the following way: "God did not leave sin alone to do its destructive work. He had and, after the fall, continued to have a purpose for his creation; he interposed common grace between sin and the creation—a grace that, while it does not inwardly renew, nevertheless restrains and compels. All that is good and true has its origin in this grace, including the good we see in fallen man. The light still does shine in the darkness. The spirit of God makes its home and works in all the creation."[27]

On this basis, he introduces a distinction between common grace and special grace:

> Even so, sin is a power, a principle, that has penetrated deeply into all forms of created life. The organism of the world itself has been affected. Left to itself,

---

23. Schreiner, *The Theater of His Glory*, 30.

24. See Gilkey, *Reaping the Whirlwind*, 179, 183.

25. I have discussed Kuyper's notion of common grace at some length in an extensive essay entitled "Abraham Kuyper's Legacy for Contemporary Ecotheology: Some Reflections from within the South African Context," 3–146. The discussion in this section partially draws on that earlier essay.

26. Kuyper, *De Gemeene Gratie* I, 243–50.

27. Bavinck, "Common Grace," 51.

sin would have made desolate and destroyed all things. But God has interposed his grace and his covenant between sin and the world. By his common grace he restrains sin with its power to dissolve and destroy. Yet common grace is not enough. It compels but it does not change; it restrains but does not conquer. Unrighteousness breaks through its fences again and again. To save the world, nothing less was needed than the immeasurable greatness of the divine power, the working of his great might which he accomplished in Christ when he raised him from the dead and made him sit at his right hand in the heavenly places (Eph. 1:19, 20). To save the world required nothing less than the fullness of his grace and the omnipotence of his love.[28]

What is at stake in this distinction between special grace and common grace is an understanding of the relationship between nature and grace and related distinctions between gospel and culture, general revelation and special revelation, general providence and special providence,[29] revelation and experience, faith and reason, science and theology, being human and being Christian, church and society, the Sabbath and the work day, and so forth.[30]

Bavinck develops this notion of common grace polemically over and against Roman Catholic, Anabaptist, and Socinian positions that were, in his view, still dominant by the late nineteenth century.[31] For him, the gospel assumes a fundamental distinction between sin and grace, not nature and grace. Grace is not opposed to nature but is in a sense itself natural (and thus common). Nature is reborn by grace: "Grace does not remain outside or above or beside nature but rather permeates and wholly renews it."[32] Likewise, God's work of salvation is not opposed to creation but is best understood as re-creation. This is contrasted with the Roman Catholic distinction between natural and supernatural religion where grace (*donum superadditum*) supplements nature so that natural knowledge of God is supplemented by revealed knowledge. Through the sacraments working *ex opere operato*, the church enables grace to infuse a human being, making

---

28. Bavinck, "Common Grace," 61.

29. It may be noted that, in Reformed orthodoxy, the distinction between general (*providentia generalis*) and special providence (*providentia specialis*) refers to God's governance in the whole world order as such, as opposed to in the church and among the faithful. A further distinction was made between ordinary and extraordinary providence, namely God's governance through the laws embedded in nature and history vis-à-vis in the form of miracles. See Barth, *Church Dogmatics III.3*, 184–85. I concur with David Fergusson that such distinctions need to be blurred somewhat as God's care is more polyphonic (but not cacophonic) than such distinctions suggest. See his *The Providence of God*, 26.

30. Bavinck, "Common Grace," 56.

31. See Bavinck, "Common Grace," 55: "At the end of our century, the divinization and vilification of man and the adoration and denigration of nature are strangely mixed together."

32. Bavinck, "Common Grace," 59.

him or her capable of good works flowing forth from the supernatural fountain of love.[33]

According to Bavinck, both the Socinians and the Anabaptists rejected the Roman Catholic harmony between the natural and the supernatural (grace), albeit in opposite ways. The Anabaptist tradition rejects nature for an otherworldly supernatural grace while the Socinians (proto-modernists) reject the supernatural and consequently deify nature and culture. Bavinck explains:

> The Socinians denigrate the *ordo supernaturalis* [supernatural order] while the Anabaptists do the same to the *ordo naturalis* [natural order]. The former criticize the central mysteries of the faith—the trinity, the incarnation, and the atonement; the latter set themselves in opposition to the natural order of affairs in family, state, and society as these are recognized by Rome. The Socinians misconstrued the *gratia specialis* and retained nothing besides nature; the Anabaptists scorn the *gratia communis* and acknowledge nothing besides grace. [...] Thus the one group was conformed to the world, while the other practiced world-flight.[34]

By contrast, Bavinck argues, Calvinism entails the reformation of the natural.[35] Special grace is not absolutely but only accidentally necessary. It became necessary only because of sin, which is itself accidental in the sense that nature is not in itself sinful so that sin does not belong to the essence of things.[36] On this basis, Bavinck explains the Reformed notion of re-creation, that grace restores nature and allows it to flourish again. Christ did come to destroy the works of the devil but also to restore the works of the Father. Separatism and world-flight (escapism) are nothing but a repudiation of the first article of the Creed.[37] Bavinck explains:

> The Christian religion does not, therefore, have the task of creating a new supernatural order of things. It does not intend to institute a totally new, heavenly kingdom such as Rome intends in the church and the Anabaptists undertook at Munster. Christianity does not introduce a single substantial foreign element into the creation. It creates no new cosmos but rather makes the cosmos new. It restores what was corrupted by sin. It atones the guilty and cures what is sick; the wounded it heals."[38]

---

33. See Bavinck, "Common Grace," 47.

34. Bavinck, "Common Grace," 53.

35. Bavinck, "Common Grace," 63.

36. Bavinck, "Common Grace," 59.

37. Bavinck, "Common Grace," 60.

38. Bavinck, "Common Grace," 61.

What, then, is the role of common grace? Let me allow Bavinck again to explain at some length:

> While it is true that the world has been corrupted by sin, it nevertheless remains the work of the Father, the Creator of heaven and earth. Of his own will he maintains it by his covenant, and by his *gratia communis* he powerfully opposes the destructive might of sin. He fills the hearts of men with nourishment and joy and does not leave himself without a witness among them. He pours out upon them numberless gifts and benefits. Families [*gezinnen*], races [*geslachten*],[39] and peoples [*volken*] he binds together with natural love and affection. He allows societies and states to spring up that the citizens might live in peace and security. Wealth and well-being he grants them that the arts and sciences can prosper. And by his revelation in nature and history he ties their hearts and consciences to the invisible, suprasensible world and awakens in them a sense of worship and virtue. The entirety of the rich life of nature and society exists thanks to God's common grace.[40]

Note how Bavinck in this way attributes a legitimate power and authority not only to the state, civil society, and family, but also to culture, science, and the arts. Likewise, Kuyper treats each of these aspects at some length in Volume 3 of *De Gemeene Gratie* and in his famous *Lectures on Calvinism*. Both Kuyper and Bavinck endorsed a pluriform society with a division of powers that are each sovereign in its own right. "Subjugation of the church by the state or of the state by the church are thus both condemned."[41] The law, the media, and universities (Kuyper established the *Vrije* Universiteit in Amsterdam) are not to be controlled by the state or the church. Bavinck does mention the role of nations and peoples (*volke*), because the order extends from the nuclear family in widening circles. He does *not*, however, include the role of race (or racial purity) in such a list of structures used by God to restrain the spread of evil in the world. For Bavinck, such order remains an abundant gift from God. It becomes embedded in the world, but it cannot be taken for granted as any continuing cosmic or social order remains dependent upon God's providence.

However, once it is suggested that common grace becomes effective through these ordinances, that this helps to restrain evil in society, that this preserves what is good and creates room for the very possibility of God's particular grace, then the attention shifts to the ordinances themselves. The question is how they are uncovered and formulated. The core of the problem lies with the way in which such ordinances are based on a notion of general revelation in a way that fails to recognize the emergence and

---

39. Van Leeuwen's translation of the Dutch "*geslachten*" as "races" is in my view an obvious and unfortunate error. It is better translated as generations in the sense of extended families, extended over several generations beyond the nuclear family (*gezin*).

40. Bavinck, "Common Grace," 59.

41. Bavinck, "Common Grace," 63.

evolution of variable social structures in history and the impact of human sin on such structures. For Kuyper and Bavinck, one may say that these ordinances are not so much embedded in the created order (*ordo*) itself but follow from the ordering (*ordinatio*) of common grace, that is, from the divine response to human sin.[42]

■ Common Grace and the Orders of Creation

There is no need here to offer a detailed discussion of the unsavory debates on German *Ordungstheologie*. These "orders" are typically understood as providential orders through and within which God accomplishes God's work of conservation. These typically include various social relationships such as marriage, the family, civil society, the state, and so on. In the German context, such orders were extended to categories of people ("*volk*") and race. At a later stage, soil (fatherland) was also added. Gradually, nation and state, race, and blood and soil became decisive. This allowed the so-called German Christians, on the basis of such an *Ordungstheologie*, to link being German and being Christian inextricably together. The logic of such a theology of creation ordinances is well captured by the South African theologian Jaap Durand:

> According to Werner Elert, who together with Emanuel Hirsch, Friedrich Gogarten and even Paul Althaus was one of the more famous ordinance theologians, two important elements can be distinguished in the idea of creation ordinances: a dispositional element and an ethical element. Both these two elements are part of God's providential order through which God continues God's work of conservation. The element of disposition points to the destiny of each person with regard to the place and position he [or she] occupies in life. The Creator places a person within a particular people, state and society. They form part of a specific race and inhabits a specific homeland. In regard to these things they do not really have a choice. This is divine providence. But at the same time it also entails an ethical obligation. Because God has decreed these things for me, I must honor and obey them. With the exception of, among others, Paul Althaus, the ordination theologians maintained that the creation ordinances were unaffected by sin. Therefore, according to them, humans can derive God's will for their lives from these ordinances, outside and apart from the message of Christ.[43]

The opposition to such *Ordungstheologie* within the confessing church movement, in which Karl Barth and Dietrich Bonhoeffer played leading

---

42. For this paragraph, see footnote 25 above.

43. Durand, *Skepping, Mens, Voorsienigheid*, 194–95 (my gender-inclusive translation). In this section, I am drawing on Durand's excellent discussion (only available in Afrikaans) of debates on providence and the orders of creation in Germany and South Africa (194–97). Durand was one of my predecessors in teaching systematic theology at the University of the Western Cape.

roles, is well-known and need not be discussed here.[44] At the first meeting of the Confessing Church at the Synod of Barmen (1934), an unequivocal position was taken against an *Ordungstheologie*. The famous first article of the Barmen Declaration reads:

> Jesus Christ, as he is attested for us in holy scripture, is the one Word of God which we have to hear and which we have to trust and obey in life and in death. We reject the false doctrine, as though the church could and would have to acknowledge as a source of its proclamation, apart from and besides this one Word of God, still other events and powers, figures and truths, as God's revelation.[45]

Nevertheless, such opposition to German *Ordungstheologie* did not lead to the rejection of every notion of the orders of creation or providential orders. Emil Brunner, for example, defended the role of the natural order and the social order (especially marriage and family) as part of God's sustaining activity to allow people alienated from each other by sin to live together again. Since the basic structure of the orders remains constant, it can be used as a principle for a social ethic. However, because a particular social order could express human wickedness, any such creation ordinances must be subordinated to the commandment of Christ.[46] Likewise, Dietrich Bonhoeffer defended the role of four "divine mandates" (or commissions), namely church, marriage and family, culture and government (thus excluding any reference to *volk*, race or soil). Yet Bonhoeffer's approach is radically different from that of the German *Ordungstheologie* insofar as he brings the four mandates into closer connection with Jesus Christ. Only in Christ, Bonhoeffer emphasizes, do all things stand and only to him are all things directed.[47] Subsequent German Reformed theologians such as

---

44. For Barth's own discussion of God's preserving, upholding, and sustaining of creatures, see his *Church Dogmatics III.3*, 58–90. For Barth, as for Calvin before him, the continuity of any creature is dependent upon God's work of preservation. It can have continuity only as God gives it (75). However, Barth then explains that the threat to such continuity is not only posed by non-being as opposed to being, but also to what was not willed and elected by God the Creator, what was excluded and rejected (76). For Barth, that is the meaning of chaos. Preservation is then preservation against the chaos of nothingness (*das Nichtige*). It therefore includes a salvific element, captured by Barth with the Latin *servare* and not only *conservare* (75). Barth offers a Christological rationale for this, namely that God preserved the creature (and then almost necessarily so) for the sake of the Son by becoming a creature (79). Barth's most detailed discussion of *das Nichtige* (the problem of nothingness) is found in the same volume of the *Church Dogmatics* on the doctrine of providence (299–368). See the very different take on chaos by Arnold van Ruler in the discussion below, developed precisely in opposition to Barth's position.

45. See https://www.ekd.de/en/The-Barmen-Declaration-303.htm [last accessed October 10, 2024].

46. See the German title of Brunner's *Das Gebot und die Ordnungen: Entwurf einer Protestantisch-theologischen Ethik* (Zwingli Verlag, 1939), translated as *The Divine Imperative: A Study in Christian Ethics* (Lutterworth, 1958). The controversy between Brunner and Barth on nature and grace, and the legitimacy of some form of natural theology is obviously relevant here. See Brunner, "Nature and Grace" and Barth, "'No!' Reply to Emil Brunner's 'Nature and Grace'." See also my discussion of this debate in "Abraham Kuyper's Legacy."

47. See Bonhoeffer, *Ethics*, 388–408.

Jürgen Moltmann and Michael Welker seem to avoid any discussion of the orders of creation in their contributions to creation theology.[48]

In the South African context, apartheid theology may be understood as a particularly crude expression of the way in which God's providence is associated with the orders of creation and a particular social order. Apartheid theology, drawing on Dutch and German sources, emerged in the 1930s as a biblical and theological legitimation of apartheid, a policy that the Dutch Reformed Church (especially) vehemently proposed to address the "poor white" question in the wake of the South African War (1899–1902) and the Great Depression. Ironically, this was meant as a critique of the impact of British imperialism but adopted and radicalized the British policy of racial segregation implemented in many of its colonies. Indeed, one may say that apartheid theology was a form of liberation theology (from British hegemony) gone wrong.

The story of the rise and fall of apartheid theology has often been told and does not need to be treated here in any further depth.[49] The categories employed to offer a legitimation of apartheid were derived mainly from neo-Calvinist sources with specific reference to Abraham Kuyper's notion of sphere sovereignty and Herman Dooyeweerd's idea of law (*wetsidee*). The intellectual foundations of apartheid theology were laid by the philosopher Henk Stoker with his notion of the idea of creation (*skeppingsidee*) and by *volksvaders* such as Totius (Jacob Daniël du Toit). Essentially, apartheid theology expanded the notion of the orders of creation to include race, ethnicity, people (*volk*), and nationality, tied with issues of language and land ownership. For Dooyeweerd, social institutions were the product of a process of differentiation in history. For Stoker, these institutions are embedded in the very structures of creation itself; they are therefore more rigid and they receive ultimate legitimacy.[50] Either way, these orders were regarded as part of God's providential order to curb the spread of evil, including conflict between ethnic groups.

The use of such categories centered around a reading of the Babel narrative (read together with Acts 17:26). Accordingly, God prevented the rise of imperialism not only through a confusion of languages but also through maintaining racial, ethnic, and cultural diversity. People have to be kept apart (through law and order) and, if necessary, even be forced apart, or else one group will begin to dominate the other. A failure to maintain

---

48. See Moltmann, *God in Creation*; Welker, *Creation and Reality*.

49. In my assessment, the best analyses are written from within and then in Afrikaans. See especially Kinghorn, *Die NG Kerk en Apartheid*, and Coetzee, *Die "Kritiese Stem."*

50. Conradie, "Creation and Salvation: Dialogue on Abraham Kuyper's Legacy for Contemporary Ecotheology."

such separation can only breed conflict. On the basis of laws on race classification that were proposed and implemented, this eventually led to a prohibition against racially mixed marriages, separate amenities, and segregated housing in "group areas" (prompting forced removals and influx control) and to an envisaged constellation of independent South African states ("Bantustans"). Such policies of separation on the basis of race were already implemented in the church itself through separate communion cups, separate worship services, separate congregations, and eventually (by 1881) separate churches.

Ethnic groups therefore needed to be kept apart, if necessary by imposing "law and order," since this was enshrined not only in historical developments (God's providence) but in the very orders of creation. In separation, later separate development, lay nothing short of "our" salvation—so that apartheid also operated as a quasi-soteriology.[51] Ironically, God's work of providence, which was aimed at preventing things from falling apart, was inverted to ensure that things were kept apart! The social order that was supposed to curb the spread of evil only exacerbated such evil—in the name of God's providence.

Apartheid theology culminated in a document entitled *Ras, Volk en Nasie en Volkereverhoudinge in die Lig van die Skrif* (1974),[52] which was accepted by the General Synod of the Dutch Reformed Church as guiding the problem of racial relations. According to this report, ethnic diversity was already embedded in God's intentions for creation and was to be appreciated as God's blessing.[53] The unity of the human race was emphasized (on the basis of the Genesis 10—because the whole human population "descended from Noah"!)[54] while any "humanistic" effort to enforce uniformity was criticized as typical of the hubris of empire building. Such a process of differentiating "gained momentum" with the confusion of tongues after Babel, eventually "deepening" to a diversity of races, ethnic groups, peoples and nations.[55] This formed part of God's providential order to prevent one group from dominating another. Diversity was therefore to be welcomed while it was sin that caused alienation and conflict between people. On the basis of this argument, the document

---

51. See Coetzee and Conradie, "Apartheid as Quasi-soteriology."

52. The document was translated into English as *Human Relations and the South African Scene in the Light of Scripture*, but the "human relations" in the title hides the use of the categories of race and *volk* and underplays the emphasis on ethnicity and nationhood.

53. Ned Geref Kerk, *Ras, Volk en Nasie*, 13.

54. Ned Geref Kerk, *Ras, Volk en Nasie*, 11. The document adopts a rather literalist reading of the Genesis narratives.

55. Ned Geref Kerk, *Ras, Volk en Nasie*, 15, 19.

offers a qualified "yes!" to the question whether a policy of separate (*eiesoortige*) development could be based on Genesis 11:1–9.[56] Ironically again, this document (written with an aura of erudite scholarship) was accepted at a time just before apartheid as the dominant political ideology was about to be replaced by an emphasis on state security following the Soweto uprisings in 1976. As a policy document, it was to be replaced by *Kerk en Samelewing* (*Church and Society*, 1986) where the category of *volk* is toned down and where separate development (a euphemism for apartheid) was no longer legitimized.[57]

The inclusion of the categories of race, ethnicity, and soil as part of the orders of creation that God presumably employed to curb the spread of evil is easy to critique given the ideologies associated with Nazism and apartheid. However, one may also pose similar questions regarding the other proposed orders of creation, given current debates on marriage (heteronormativity), the family (patriarchy), work (in a capitalist global economy), government (the dangers of tyranny), and the state (given the dangers of "state theology,"[58] the ideology of state security and state capture). The same could apply to organizations in civil society (including churches), although the church can hardly be regarded under the rubric of God's providence. Moreover, each of these human institutions emerged during the course of human history so that such orders cannot be embedded in God's work of creation. They are situated within particular cultural contexts that are subject to change so that any rigid sense of divine order can hardly do justice to a God of history (hence *gubernatio* more than *conservatio*).

An almost deterministic emphasis on God's sovereign will, that everything is ordained by the will of God, cannot do justice to the biblical narratives that more typically suggest that the world is not the way God wants it to be. This should incite complaint and protest, engagement in social transformation, not a resignation to injustice or a placid acceptance that all suffering comes from the hand of God.[59]

This is also the avenue followed by Jürgen Moltmann in *Ethics of Hope*. He affirms the Protestant sense of responsibility for the world. At times, the affirmation of the world implies the need to change unjust structures—so that swords are turned into ploughshares. Moltmann does not recognize social and political structures as God's ordinances but as human constructions for which humans have to take responsibility.[60]

---

56. Ned Geref Kerk, *Ras, Volk en Nasie*, 16, 72.

57. See https://kerkargief.co.za/doks/bely/GD_KerkSamelewing.pdf [last accessed October 10, 2024].

58. See the critique of state theology in the *Kairos Document* (1986).

59. See, Fergusson, *The Providence of God*, 7.

60. See Moltmann, *Ethics of Hope*, 206.

■ A God of Order in Nature and Society or a God Transforming Society and Nature?

This emphasis on the orders of creation and divine ordinances raises the question whether it may not be best to discard all attempts to defend such orders of creation as a sign of God's providence. Since such orders form a crucial part of "common grace," should such a theological construct not be discarded as well? Many may be inclined to suggest that any notion of God's providence needs to be abandoned as hopelessly entangled in the legitimation of power. In the words of Susan Neiman, "Providence is a tool invented by the rich to lull those whom they oppress into silent endurance. The rich have no need of virtue or faith for their desires are met without them. [...] Providence is either a tool invented for oppression or itself an instrument of injustice."[61] Is it not safer, instead, to focus on the gospel of liberation from evil, forgiveness of sins for the sake of reconciliation with God, and the transformation of society? Or is it even better to cut God out of the picture altogether?!

It is important to recognize the full extent of what is at stake here. Under the influence of Greek assumptions on stable forms in nature and equally stable social structures in history, God's providence was typically understood in terms of preserving order in nature and law and order in history, therefore as *conservatio*. Accordingly, providence is equated with the preservation of creatures over time.[62] Religion thus served the purpose of sacralizing power and divinely ordained structures and not as instigating social change. God's work can be understood as maintenance and restoration, perhaps as reform but not as revolution.[63] If anything, the city of God remained suprahistorical.

One may say that Joachim of Fiore's notion of distinct epochs in history, a Lutheran sense of vocation, and a Calvinist emphasis on sanctification[64] each recognized the significance of social transformation, but it was with the advent of modernity that the emphasis shifted not only from *conservatio* to *gubernatio* but also from maintaining order to social transformation in history, to the dream of building a better society, even a classless utopia. With the emergence of a historical consciousness, social structures could

---

61. Neiman, *Evil in Modern Thought*, 182.

62. See Gilkey, *Reaping the Whirlwind*, 250.

63. See Gilkey, *Reaping the Whirlwind*, 192.

64. There is, for Calvin as for Augustine, a transcendent, suprahistorical, eschatological goal, understood as being with God, on earth as it is in heaven. But, unlike Augustine, Calvin also emphasized the regenerative impact of grace on life and therefore on history. Sanctification is not based on justification, but the aim of justification is sanctification. The process of sanctification therefore represents the transformation of the lives of believers in the world and through them of the character of the world. See Gilkey, *Reaping the Whirlwind*, 184.

no longer be regarded as divinely ordained forever but as contingent, culturally relative, and subject to change. However, the vision was no longer based on divine intervention in history but on scientific progress and technological innovation, on market incentives, on corporate risk-taking, and on the internal democratic impulse for freedom, equality, and solidarity. This led to a desacralization of history and an increasingly radical critique of God's providence as the enemy of any social transformation.[65]

In response to such modernist views on progress, the kingdom of God was reconceptualized as a vision for the future instead of something suprahistorical.[66] God is then not the God of the present who has created and upholds the present order of things.[67] God, the God of becoming and not being, becomes the lure of the future that draws us towards an attractive vision for society in an evolutionary way towards a divinely ordained end (as assumed in nineteenth-century liberal theologies). If anything, conservative notions of providence as conservation are expelled and replaced with providence as development and progress.[68] However, if and when that vision fails, when progress itself is equated with ecological gloom and doom, then God is portrayed as standing in radical opposition to the evil present, calling for social transformation, promising a new dispensation in opposition to the old (as assumed in political and liberation theologies since the 1960s).[69] God is not the God of the present who is above us, behind us, with us, and in us as the sovereign Lord.[70] Such a God would

---

65. See Gilkey, *Reaping the Whirlwind*, 199.

66. See the following comment by Jaap Durand: "While in the preceding liberal theology the developing history in nature and human society was seen as the place of God's saving actions and the future completion of human history as God's eschatological end goal, in the new crisis theology nature and history become detached from God's saving presence. As a result, the eschatological end goal of God and the future of the human community are also disconnected. The theological effect of this on the doctrine of providence was inevitable. As a doctrine within theology it has lost its relevance" (my translation). See Durand, *Skepping, Mens, Voorsienigheid*, 199.

67. See Gilkey, *Reaping the Whirlwind*, 229.

68. See Gilkey, *Reaping the Whirlwind*, 210.

69. Gilkey explains the significance of such a historicizing of eschatology in a lucid way, worth quoting at some length: "The final eschatological purposes of God, [...] are no longer transhistorical, intending a salvation for individuals beyond history in eternity [as with Augustine and Calvin]. On the contrary, these eschatological purposes now concern the historical itself, the end of history in the literal sense, since that end is now conceived to be the perfection of the humanum, a concrete historical community of justice, peace, freedom and communion. The symbol of the kingdom becomes a symbol for the character of future social history; and the divine activity in history—and so the major task of the church—is seen as an activity which prepares that culmination by working, not simply in the church and on individual souls, but in the social process itself in the development of democratic rights, economic justice, social equality and international peace. As the eschatological goal was thus moved into historical process itself, so God's work in the world, and a Christian obligation toward the world, was likewise moved into 'secular' historical process to transform that process creatively." See Gilkey, *Reaping the Whirlwind*, 202–03.

70. See Gilkey, *Reaping the Whirlwind*, 231.

ordain the evils of history and would be the enemy of human freedom. Instead, God becomes the ultimate power of the future, impinging on the present to negate it and transform it. The eschatological vision is not developmental or progressivist but revolutionary through a radical negation of the present.[71] The vision is for a new creation that would overcome both social evil and natural evil.

It then comes as no surprise that any notion of God's providence, not only of maintaining order in nature and in society but also of governing history, becomes increasingly sidelined (if not vanishing completely)[72] and subsumed under an emphasis on God's work of salvation (better: liberation) and consummation. Moreover, liberation and consummation may then be regarded as immanent forces so that God might as well be left out of the picture altogether. This heightened the problem of *concursus*, but the question was no longer one of making room for human freedom over against divine determinism but to explain God's activity in social processes if history as a whole is understood in terms of immanent material and social forces.[73]

■ Sustainability and Holocene Stability

There is some irony here, namely that secular society can hardly abandon a secularized form of providence understood as *conservatio*. This becomes evident from the role played by the natural order and the social order in contemporary discourse on sustainability.

In recent contributions, I noted that diverse terms are employed in contestation with each other in order to speak of a sustainable society, sustainable growth, sustainable development, sustainable communities, and sustainable livelihoods.[74] I reviewed the shifts and turns in such secular discourse on sustainability. In short, these shifts include an initial focus on nature conservation and wilderness preservation, the emerging concern over limits to the use of nonrenewable resources given issues of population and consumption (the report on *Limits to Growth*), a focus on the sustained use of renewable resources (the report on *Our Common Future*), a concern over the destructive social impact of environmental degradation (e.g., environmental racism, at least since the Rio Earth Summit), a recognition of

---

71. See Gilkey, *Reaping the Whirlwind*, 229–30.

72. See Gilkey, *Reaping the Whirlwind*, 231, arguing that God's providence in history became even more problematic in the eschatological theologies of the 1960s than in the neo-orthodox theologies of the 1920s.

73. See Gilkey, *Reaping the Whirlwind*, 199.

74. See my *Secular Discourse on Sin in the Anthropocene*; also "What, Exactly, Needs to Be Sustained amidst a Changing Climate?"

the limits to biophysical throughput measured in terms of an environmental footprint (carrying capacity), an increasing awareness of biospheric limits to absorb the waste products of an industrialized economy (absorption capacity), most notably greenhouse gases, and finally discourse on planetary boundaries and thresholds. In such debates, some focus on the question *whether* the industrialized global economy can be sustained (or will it collapse), some on *how* it can be sustained (through appropriate technologies) and for *how long* it can be sustained (scenario-planning taking into account the unequal use of resources, causing unequal environmental impact and with an unequal distribution of costs).

The deeper question, though, revolves around *what* it is that is to be sustained. The current use of resources? The unequal use of resources, perhaps? Economic growth in terms of biophysical throughput? Socio-economic development based on such growth? Or, to be more honest, a consumer-driven middle-class lifestyle? Or Western notions of civilization? Or being brutally honest: industrialized neoliberal capitalism? Or more soberly: what is to be sustained are treasured institutions and events despite their environmental footprint (e.g., universities, mega-sports events, tourism), which would require sacrifices to be made elsewhere in the system. With discourse on planetary boundaries, it may be best to answer this question with reference to Holocene stability. The various planetary thresholds need to be guarded in order to protect the relative stability that characterized the Holocene epoch within which agriculture became possible, within which all the major civilizations emerged, and within which family life, culture, religion, the arts, and sciences could flourish. This is, I would suggest, the bottom line in discourse on sustainability.

It is remarkable to note how the social (or economic) order is related to the natural order in such discourse on sustainability. Some may emphasize what is needed in terms of the social order (science, innovative technologies, increasing prosperity, development) and ask how this is possible within planetary boundaries. Others take the balance within the Earth System (in the singular but with various subsystems) as the point of departure in order to argue that the global economy needs to stay within such planetary boundaries, bringing deleterious socio-economic development under control (which may well be wishful thinking). But in one way or another, there is a recognition that the social order and the natural order need to be compatible with each other—for the sake of Holocene stability. To be more precise, there is a need to understand the interplay between the Earth's slow-moving lithic strata and the faster-flowing thin envelope of water, air, and life (and social orders within that) around the Earth's surface.[75]

But what if the Holocene is already no more? The proposed date for the advent of the so-called "Anthropocene" (to be marked by a "golden spike") is by the mid-twentieth century.[76] If so, we have already been living within the "Anthropocene" for three generations. Then it makes no sense to preserve Holocene stability, although it would be possible to slow down the process or at least to manage the transition and the tumultuous changes associated with that. The question is whether the concept of sustainability (and adaptability to cope with any deviations) is appropriate to serve such purposes. Or should it be discarded together with nineteenth-century notions of progress, twentieth-century assumptions about sustained economic growth, and twenty-first-century discourse on sustainable development goals (SDGs)? Or is the problem that not only sustainability but also adaptability is being degraded, not only the adaptability of human societies but the Earth's own capacity for self-differentiation, allowing for the generation of novelty?[77]

■ Ecojustice and the Need to Disrupt an Unjust Global Order[78]

The theme of justice has long been on ecumenical agendas. It is captured in the motto "Towards a Just, Participatory and Sustainable Society" from the World Council of Churches' Nairobi Assembly (1975). It is again expressed in the "conciliar process" towards "Justice, Peace and the Integrity of Creation" initiated at the Vancouver Assembly (1983) and in the conference theme "God of Life, Lead Us to Justice and Peace" of the Busan Assembly (2013). Justice is understood in a comprehensive way with the connotations of social justice (e.g., the Program to Combat Racism), economic justice (e.g., the Accra Confession of 2004 and the document on *Alternative Globalization for People and Environment—AGAPE*, 2005), gendered justice (e.g., the Decade of Churches in Solidarity with Women and the Decade to Overcome Violence), and then also environmental justice, and more specifically climate justice.

---

75. See Clark and Szerszynski, *Planetary Social Thought*, 64.

76. See the homepage of the Anthropocene Working Group at http://quaternary.stratigraphy.org/working-groups/anthropocene/ [last accessed October 10, 2024]. For a discussion, see also my "Some Theological Reflections on Multi-disciplinary Discourse on the 'Anthropocene'."

77. See Clark and Szerszynski, *Planetary Social Thought*, 174.

78. This section partly contains elements of and builds upon my recent article, "The Limits of Ecological Justice?"

The term "ecojustice" is often used in ecumenical discourse to capture the need for a comprehensive sense of justice that can respond to economic injustice, ecological degradation, *and* the interplay between them. This term was popularized by William Gibson and Dieter Hessel.[79] It builds upon the recognition that the English words ecology, economy, and ecumenical share the same etymological root in the Greek *oikos* (household). Accordingly, ecology describes the underlying logic (*logos*) of the household, economy circumscribes the rules (*nomoi*) for the management of the household, while the "whole inhabited world" (*oikoumene*) refers to the (human) inhabitation of the household.[80] This implies that "nature" cannot be regarded as a dimension of the economy (as "resources" or "externalities," as assumed in environmental economics); the economy is situated within and forms part of the natural world (as assumed in ecological economics). Ecojustice therefore encompasses more than environmental justice while certainly including that.

Given the above, the need to extend notions of justice to climate justice became self-evident in ecumenical discourse on climate change. One may observe that the principle of climate justice is widely understood and affirmed, namely that those nations that will be most adversely affected by the impact of climate change contributed relatively little to historical carbon emissions. The focus is on those countries that are particularly vulnerable to the adverse effects of the climate change given rising sea-levels, droughts, floods, and other disasters. This is epitomized by Small Island States such as Tuvalu and Kiribati, but most African countries would be adversely affected by climate change in a way that is inversely proportioned to historical carbon emissions. South Africa's position is somewhat different in that it is still thirteenth on the list of carbon emissions by country (despite load shedding!),[81] even though the South African government is posturing to ensure access to such climate funds in order to meet its mitigation targets.[82] An energy transition is clearly needed, but, as

---

79. See Dieter Hessel, *After Nature's Revolt*.

80. See, e.g., Rasmussen, *Earth Community, Earth Ethics*. The whole household of God has been widely discussed in countries such as South Africa and South Korea and throughout Pasifika. The Pacific Conference of Churches reconceptualized ecumenism as a "household" in 2018. See Vaai and Casimira, *reSTORYing the Pasifika Household*. Nevertheless, the *oikoumene* then has to be interpreted without an imperial or kyriarchal homogeneity.

81. There are several such lists available on the internet. For one example, see https://www.worldeconomics.com/Indicator-Data/ESG/Environment/Carbon-Emissions/ [last accessed October 10, 2024].

82. In *A Framework for a Just Transition in South Africa* the Presidential Climate Commission reports that "Achieving a just transition in South Africa will require significant capital mobilisation, from both public and private sources, both domestically and internationally. It is estimated that South Africa will require at least US$250 billion over the next three decades to transform the energy system 2022." (24).

Pope Francis also notes, this would need to meet three conditions: that measures be efficient, obligatory, and readily monitored.[83]

However, the call for ecojustice should not be domesticated, as if this is merely an addendum to discourse on sustainability and more specifically sustainable development. If so, climate mitigation has precedence in the Global North, while climate adaptation is on the agenda especially of the Global South, that is, of the victims of climate change. Technology transfer and a climate loss and damage fund are then added to this agenda for the sake of a degree of justice. What is at stake here is a more fundamental critique of the current global economic order. The debate is not so much on the role of a free market, the regulation of the market or a state-controlled economy. It is the core assumption of the need for sustained economic growth as expressed in the form of globalized neoliberal capitalism that is questioned.

There is a wealth of ecumenical literature where such a critique is expressed. It may suffice to quote several sections from the Accra Confession (2004) in this regard:

> 8. The policy of unlimited growth among industrialized countries and the drive for profit of transnational corporations have plundered the earth and severely damaged the environment. In 1989, one species disappeared each day and by 2000 it was one every hour. Climate change, the depletion of fish stocks, deforestation, soil erosion, and threats to fresh water are among the devastating consequences. Communities are disrupted, livelihoods are lost, coastal regions and Pacific islands are threatened with inundation, and storms increase. High levels of radioactivity threaten health and ecology. Life forms and cultural knowledge are being patented for financial gain.
>
> 9. This crisis is directly related to the development of neoliberal economic globalization, which is based on the following beliefs:
> - unrestrained competition, consumerism and the unlimited economic growth and accumulation of wealth are the best for the whole world;
> - the ownership of private property has no social obligation;
> - capital speculation, liberalization and deregulation of the market, privatization of public utilities and national resources, unrestricted access for foreign investments and imports, lower taxes and the unrestricted movement of capital will achieve wealth for all;
> - social obligations, protection of the poor and the weak, trade unions, and relationships between people, are subordinate to the processes of economic growth and capital accumulation.
>
> 10. This is an ideology that claims to be without alternative, demanding an endless flow of sacrifices from the poor and creation. It makes the false promise that it can save the world through the creation of wealth and prosperity, claiming sovereignty over life and demanding total allegiance which amounts to idolatry.

---

83. Pope Francis, *Laudate Deum*, §59.

And:

> 17. We believe in God, Creator and Sustainer of all life, who calls us as partners in the creation and redemption of the world. We live under the promise that Jesus Christ came so that all might have life in fullness (Jn 10.10). Guided and upheld by the Holy Spirit we open ourselves to the reality of our world.
>
> 18. We believe that God is sovereign over all creation. "The earth is the Lord's and the fullness thereof" (Ps 24.1).
>
> 19. Therefore, we reject the current world economic order imposed by global neoliberal capitalism and any other economic system, including absolute planned economies, which defy God's covenant by excluding the poor, the vulnerable and the whole of creation from the fullness of life. We reject any claim of economic, political and military empire which subverts God's sovereignty over life and acts contrary to God's just rule.
>
> 20. We believe that God has made a covenant with all of creation (Gen 9.8–12). God has brought into being an earth community based on the vision of justice and peace. The covenant is a gift of grace that is not for sale in the market place (Is 55.1). It is an economy of grace for the household of all of creation. Jesus shows that this is an inclusive covenant in which the poor and marginalized are preferential partners and calls us to put justice for the "least of these" (Mt 25.40) at the centre of the community of life. All creation is blessed and included in this covenant (Hos 2.18ff).
>
> 21. Therefore we reject the culture of rampant consumerism and the competitive greed and selfishness of the neoliberal global market system or any other system which claims there is no alternative.
>
> 22. We believe that any economy of the household of life given to us by God's covenant to sustain life is accountable to God. We believe the economy exists to serve the dignity and wellbeing of people in community, within the bounds of the sustainability of creation. We believe that human beings are called to choose God over Mammon and that confessing our faith is an act of obedience.
>
> 23. Therefore we reject the unregulated accumulation of wealth and limitless growth that has already cost the lives of millions and destroyed much of God's creation.
>
> 24. We believe that God is a God of justice. In a world of corruption, exploitation and greed, God is in a special way the God of the destitute, the poor, the exploited, the wronged and the abused (Ps 146.7–9). God calls for just relationships with all creation.
>
> 25. Therefore we reject any ideology or economic regime that puts profits before people, does not care for all creation and privatizes those gifts of God meant for all. We reject any teaching which justifies those who support, or fail to resist, such an ideology in the name of the gospel.

One may therefore say, if pushed to the extreme, that an emphasis on sustainability (especially in the form of sustainable development) and on ecojustice represent two contrasting social visions. The one seeks to make the current global economic order more sustainable while the other rejects that order and calls for a transformation of that order. There may be at times a need to establish order amid reigning chaos, but (as the South African experience demonstrates) there may also come a time when it is necessary to disrupt an unjust social order in order to construct (or reconstruct) a new social order where justice and peace can reign. Ironically, at times, the need for "development" is recognized especially in the Global South while a critique of the global economic order is also voiced in the Global North.

■ Van Ruler on God and Chaos

How, then, should one steer between Holocene stability and "Anthropocene" volatility, between the inequalities of a neoliberal order and the chaos of collapsing markets? How should one think theologically about that? Is God's providence always to be understood in terms of a natural and a social order that can restrain and contain evil? Is chaos a sign of evil that needs to be curbed (as Calvin took for granted)? How does one prevent a social order from becoming hegemonic and from legitimizing structural violence? Where does the energy come from to disrupt and transform an oppressive social order? Is there not a need for any form to be reabsorbed into the formless (the void, chaos, orgy, water, the subliminal) to recover its vigor?[84] Or does the modernist triumph of *nous* prevailing over chaos lie at the very root of ecological destruction?

Remarkably, some assistance in this regard may come from a follower of Herman Bavinck, namely Arnold van Ruler. As far as I could establish, Van Ruler did not use the term common grace in his mature work,[85] although he wrote extensively on nature and on grace.[86] He did affirm Bavinck's position that the fundamental distinction is between sin and grace but avoided the term grace in relation to nature (understood as creation) itself. God's grace is aimed at overcoming the problem of human sin, understood primarily as guilt. For Van Ruler, salvation is not an aim in itself; it is not about the experience of salvation, or about being saved, or even about the Savior, but about the being that is saved. Salvation is necessary in order to allow God's creation to be, to exist before God's face.[87] Put differently, grace is there for

---

84. See Eliade, *Cosmos and History*, 88.

85. Van Ruler did discuss common grace in an early work on Abraham Kuyper's idea of a Christian culture (*Kuyper's Idee eener christelijke Cultuur*). His argument there is that Christianizing culture (Niebuhr would say "Christ transforming culture") is necessary because of sin but that the point is not that culture should become Christian but that Christianity will enable culture to flourish. Accordingly, Kuyper was not radical enough in his vision.

86. The term common grace is not used in any form in Volume 3 of Van Ruler's *Verzameld Werk* that collects his writings on God's providence.

87. For a discussion, see my essay, "Van Ruler as an Early Exponent of Christian Ecotheology?"

the sake of nature, the gospel for the sake of culture, Christianity for the sake of the world, the church for the sake of society. For Van Ruler, paying taxes, insofar as its aim is to redistribute wealth, is even holier than the Holy Communion.

In a famous, hotly debated essay on God's providence entitled "God and Chaos," Van Ruler argues that chaos forms part of God's good creation. He resists any Manichean notion that chaos could be co-original with God but also any Barthian notion that chaos emerged from the suction power of nothingness (*das Nichtige*).[88] Instead, he suggests that *God* created chaos to play with it. Leviathan, the prime symbol of chaos was created by God to play with it (as per Ps 104:26). Van Ruler adds that, "God does not fear chaos. God does not want to be rid of it as soon as possible. God *plays* with it. All who play become engrossed in the play. There is something of eternity in play. Could it be that from eternity to eternity, God plays *with* and becomes engrossed *in* the chaos?"[89]

What does it mean that chaos is the work of God? Van Ruler explains:

> But the earth was waste and void in the beginning. There, then, chaos was the creative work of God. It is even the first thing that emerges. The Lord God does not shy away from it. Nor is it so that God remains caught in it. God goes to work immediately. To transform chaos into cosmos? That I don't believe. It would be better to say: to provide some fixed points within the chaos here, there and everywhere, and so to bring about a degree of order. The order can perhaps be called the midpoint between chaos and cosmos. The creative God is permanently engaged in ordering. That is, on the one hand, forming, sculpting, giving profile [*gestalte*]; on the other hand, it is, however, also casting out, separating, sifting and judging. We could also say: all divine ordering is both restraining of the chaos and breaking up of the cosmos.[90]

Moreover, it is not as if God merely allows chaos or is somehow threatened by it. It is precisely Godself who chaotifies things from time to time. God is therefore the source of chaos. Van Ruler explains this by inverting the Babel narrative:

> We see that in the Bible as well. When human beings organize life too rigidly, the living God proceeds to chaotify it. Surely that is what happened with the tower building and the confusion of tongues at Babel. There the organizing activity emanates from human beings and the chaotifying from God. In this I would even find the point of the narrative. They build the tower not to penetrate into heaven,

---

88. See also Moltmann's controversial use of the notion of a divine self-withdrawal (*zimsum*) whereby God makes room for creation but also for an annihilating Nothingness to emerge. For Moltmann, creation out of nothing therefore needs to be supplemented by the "creation of salvation" (to overcome disaster) and an eschatological new creation. Nothingness therefore remains a looming threat to creation. See Moltmann, *God in Creation*, 86–93.

89. Van Ruler, *This Earthly Life Matters*, 133.

90. Van Ruler, *This Earthly Life Matters*, 132.

but to keep the people together in a strict unity. The organization, the *state*, is superimposed like a copper dome over variegated human life in its multiplicity and variety. No room is left—at least no playroom. Therefore God intervenes. God confuses speech and thereby all co-existence and cooperation. God brings into being the chaos of a multiplicity of peoples [*volken*].[91]

On this basis, Van Ruler challenges traditional views that associate God with order (cosmos) rather than chaos:

> The living God we know from the Bible clearly does not desire true cosmos. That would be the contrary of chaos: the fully finished figure [*gestalte*], the pleasingly completed whole, the perfect balance, the pure harmony, the closed unity that is sufficient unto itself. All this the God of the Bible does not desire. As soon as people try to attain this cosmos, God acts in person or God generates God's instruments to pass through it in a chaotifying manner. Politically speaking, that holds off all strivings towards the absolute state of wholeness. Therefore, we should, on God's authority, give preference to democracy with its typical chaos elements rather than to the cosmos of communism or any other dictatorship.[92]

Van Ruler links the presence of chaos with the multiplicity of creatures. He says: "But in creation there is multiplicity. God does not create a *God*, a new unity. To think that would in itself be nonsense. In that God creates, God creates a creation. Multiplicity is inherent in that. Each creature is finite and limited. It is a fragment and a fracture. Therefore, it calls out for other creatures: they need one another in their fragmentary existence."[93]

He adds: "Who, however, would regret that? Then we would have to regret creation itself and then we step outside the Christian faith. The Bible teaches us to consistently affirm creation, joyfully and whole-heartedly—and thereby also the multiplicity of creatures and the individuation and loneliness that comes with that. The loneliness is even the fertile ground for all [sorts of] creativity. The element of chaos, or at least the possibility of chaos, that inheres in creation we can also but affirm or at least accept in a positive spirit."[94]

Ultimately, then, God allows for an interplay between chaos and order. This is arguably also found in Godself:

> God is one. In that there is no chaos. Still, there we have to be careful as well. We can, for instance, say that God is "the unity" and not at all that God is "the oneness." God is also Triune. God is one only in this mode. Thus there is also life, community, movement, in a sense multiplicity or at least threeness in God. Nevertheless, God is not chaos and there is nothing chaotic in God. God is,

---

91. Van Ruler, *This Earthly Life Matters*, 131.

92. Van Ruler, *This Earthly Life Matters*, 131.

93. Van Ruler, *This Earthly Life Matters*, 134.

94. Van Ruler, *This Earthly Life Matters*, 135.

however, not the cosmos either. God is the order itself, the *community* of the Father, the Son and the Holy Spirit.[95]

Let me leave Van Ruler's provocations aside at this point. He continues to discuss the interplay between chaos and sin (the assumption that it is humanity that chaotifies). He concludes that it is love that keeps us tied to chaos but that chaos does not paralyze love. In the context of love, order need not impose rigidity but precisely elicits creativity. Van Ruler proposes participation in a perichoretic dance. He says:

> [W]e are chaos. We are Being. We are God's play. It all depends on not only the courage to be, but far more on the willingness to be in the game. "May I dance with you?," God asks us, and the core of our being depends on the question whether we are prepared to accede. But Being as divine play, as play of divine love, is an illumination of Being, also of the chaos in Being, which would be able to satisfy and gladden even reason and the heart. That, however, demands an extensive maturity and a vast silence before the face of God.[96]

That aside, the question remains how God's work of providence, in the form of *conservatio*, could be understood in the context of the shift from Holocene stability to "Anthropocene" volatility?

■ Is It God Who Is Stirring the Soup?

The metaphor of God stirring the soup is derived from Bram van de Beek, another Dutch Reformed scholar and a one-time student sitting in Van Ruler's classes in Utrecht. In an article entitled *"Rust is Ver te Zoeken"* ("Rest is hard to come by"), partly building on Van Ruler, Van de Beek comments that "God is continuously stirring the soup. Otherwise everything will come to a standstill. And then the whole lot will be burnt. The movement, the flow has to be maintained."[97] Tectonic plate movement, as well as evolutionary movement, with all the pain and suffering it implies, is necessary for the survival of ecosystems. The Earth is a restless, dissipative planet, held far from thermal and geochemical equilibrium by its own hot interior and energy from the Sun, allowing for "irruptions of novelty" but

---

95. Van Ruler, *This Earthly Life Matters*, 134.

96. Van Ruler, *This Earthly Life Matters*, 144. From a Pasifika perspective, Upolu Vaai recommends a cosmic extension of the metaphor of dance: "It allows us to rediscover God already dancing with us to the rhythms of life, expressed in little practices such as fishing, planting, oral stories, feasting, and birthing to name a few. A theological dance informed by the silent whispers of the *vanua* (land) and the graceful movements of *vaitafe* (flowing rivers), transformed by the fluidity and unpredictability of the *moana* (ocean), animated by the *mānava ola* (breath of life) of the *vaomatua* (the elder forest), and dirtified by the rising dust from the *malae*, the ceremonial grounds of the Pacific dirt communities. A dance fused with oral stories, theologies, music, laughter, art and poetry from the village fields, and replenished by the waters and smoke of earth rituals." See Vaai, "From Atutasi to Atulasi," 236–37. Van Ruler would surely approve.

97. Van de Beek, *"Rust is Ver te Zoeken"* (my translation), 47.

also "cascades of irrevocable loss."[98] The monological tone of the geological timescale cannot do justice to the Earth's dynamic self-differentiating processes. Such processes do not exist in time but in a way "make" or "yield" time.[99] While the stasis of a sterile order may be appropriate from a deist perspective, the presence of the Spirit suggests movement and dynamism. As Langdon Gilkey observes:

> Historical change, moreover, reveals and exacerbates evils long present yet possibly hidden. By shifting the balance of power and so uncovering the oppression and injustice hidden by that stability, change brings to the surface both the decay of life and the injustice laying back of that decay. [...] From the point of view of faith, it is the strange face of the hidden God constituting, upsetting, destroying, challenging, judging, re-creating and calling.[100]

Is this how the shift from the Holocene to "Anthropocene" is to be understood? Indeed, what on earth may God be up to, given disruptions at such a scale? As with earthquakes, volcanoes, and tsunamis, this would be turbulent, going together with massive suffering and death, possibly mass extinctions, perhaps human extinction, at least with widespread disruptions in the habitat of species and the structure of human societies. The mere possibility evokes anxiety, indignation, and anticipatory grief, not least among young people.[101]

To see God's presence in our times in this way may attest to God's power and judgment but hardly to God's mercy. Indeed, the image of a God who is stirring the soup is a horrible and horrifying one, especially for the weak and vulnerable. This image is more at home in Manicheism than in a Christian understanding of God. In response, others see God not as causing chaos but as grieving profoundly at the suffering that comes with such turbulence. But does such an understanding of God do justice to the God of the Bible who also judges and condemns?[102] And is this compatible with the need to disrupt unjust social orders? What about God's work of liberation? How is

---

98. Clark and Szerszynski introduce the notion of "planetary multiplicity" to capture this process of self-differentiation on all temporal and spatial scales. For these formulations, see their *Planetary Social Thought*, 172, 173.

99. See Clark and Szerszynski, *Planetary Social Thought*, 173.

100. See Gilkey, *Reaping the Whirlwind*, 33, 34.

101. See Clark and Szerszynski, *Planetary Social Thought*, 174.

102. Van de Beek comments: "Is this not more typical of what we experience from God? We experience chaos. We experience things that he cannot resist. We experience the brokenness of human lives. We experience illness. We experience those who are exploited. Does God have nothing to do with that? Is God always only compassionate or helpful?" (my translation). See "*Rust is Ver te Zoeken*," 44–45.

mercy related to justice?[103] Who, then, is this God and what is this God doing at this moment in history?

How, then, is God's providence, common grace, and God's work of conservation to be understood in such a context? To return to the question raised in this volume: "How could the suffering of God's creatures in the 'Anthropocene' be reconciled with trust in God's loving care?" Clearly, an apophatic response to such a question is more appropriate than clever answers that cannot console those who put their trust in God's care—which remains at the heart of any adequate notion of providence.

■ Theses on God's Work of Conservation

Allow me nevertheless to conclude with ten tentative theses regarding God's work of providence if focused on conservation:

1. Despite the considerable evidence to the contrary, trusting in God's care forms a core part of the Christian faith, cannot be abandoned without abandoning faith in Godself, and is precisely epitomized by the "nevertheless" of Habakuk 3:17.[104] Providence is about faith in God's faithfulness to the whole of creation in general and God's loyalty to God's covenant partner in particular. Such merciful covenantal loyalty (חֶסֶד) is a core theme in the Old Testament and is far more descriptive of God's character than omnipotence or omniscience may be.

2. God's providence cannot be subsumed under God's work of creation,[105] or ongoing creation, or salvation, or consummation, or election, without

---

103. In literature from within the Reformed tradition, see especially Reinhold Niebuhr's *Love and Justice* and Nicholas Wolterstorff's *Justice in Love*. Niebuhr maintained that love is a truly ultimate concept, with justice its approximation under conditions of sin so that justice instead of agapeic love is called for in cases of conflict. However, Wolterstorff argued that the conflict between justice and love is a false one, proposing instead an understanding of love that incorporates justice.

104. Christian hope retains a sense of a "nevertheless" and eschews both pessimism and optimism. The scenario sketched in Habakuk 3:17 is more daunting than any the Intergovernmental Panel on Climate Change (IPCC) may have come up with: "Though the fig tree does not blossom, nor there be fruit on the vines; the yield of the olive tree fails, and the cultivated fields do not yield food; the flock is cut off from the animal pen, and there is no cattle in the stalls." But then the response is truly unheard of: "Yet I will rejoice in Yahweh; I will exult in the God of my salvation" (NRSVUE).

105. Despite Langdon Gilkey's tendency (following Whitehead and Tillich) to redescribe Christian convictions in abstract categories that do not do justice to their particularity, my sense is that his description of providence in terms of the polarities of destiny and freedom, and in terms of achieved actuality and future possibility, is on the right track. For Gilkey, God is the creative ground and ultimate goal of creaturely existence and therefore of history: "God is the power of being that carries forward the total destiny of the past into the present where it is actualized by freedom." See *Reaping the Whirlwind*, 249. As the ground of being, God transcends the transience, mortality, and contingency of creatures. But God is also the ground of possibility and therefore of autonomous human creativity in history (thus of *concursus*) (250). This enables Gilkey to see God as source not so much of conservation but of creaturely continuity and meaningful movement towards an open future (251).

messing up the story of God's economy. However, when it becomes an independent interest, it has no credibility. This is what also plagues discourse on the theodicy problem. Neither fusion nor separation is a viable option. Traces of God's providence can best be identified when a Christ-like redemptive shape is evident. So God's care and God's work of salvation cannot be separated; God's care will always have a salvific intent. However, to fuse the two is to restrict God's care to acts of salvation. Then the planetary width of God's care is lost, caring that includes the wicked, people of other living faiths, and other forms of life. The intuition that God's providence makes room for the story to continue is therefore appropriate.
3. The distinction between common grace and special grace prompts disturbing questions around the meaning of divine election and can only be retained if it becomes clear that special grace is a divine strategy for the sake of common grace so that Israel becomes a light for all nations, indeed all of creation. The Reformed intuition is that grace is there for the sake of nature so that grace is not somehow "higher" than nature. Nature need not be enriched (or "elevated") by grace. Likewise, God's special grace for the church is there for the sake of the whole world. Such a divine strategy may be risky, but the logic is clear, namely that the universal scope of God's work is based on its particularity.
4. God's providence necessarily includes the dimension of continuing creation (the constitution, sustenance, and preservation of creatures and their creativity over time)[106] but also conservation in order to restrain the spread of evil, in the same way that wilderness preservation is needed to guard against encroaching agriculture, mining, industry, and urbanization. It helps to contain environmental impact in such designated areas, but that does not by itself address the roots of the problem in economic centers of power (not in wilderness areas).
5. God's ways of curbing evil cannot be understood only in terms of conservation; the rest of the story needs to be told as well. By itself, preservation does not do justice to evolutionary history, has to address the reality that creation no longer has integrity (given the curse of Gen 3:17), and fails to express the eschatological vision that God will make all things new (Rev 21:5).[107]
6. Conservation is an always inadequate term and can be retained only if it is made clear that this could at times imply establishing order amid chaos, but at times it could also merely mean maintaining such order, or disturbing an unjust order, causing some creative chaos, or transforming any current social order for the sake of justice. Neither the triumph of

---

106. See also Gilkey, *Reaping the Whirlwind*, 272.

107. See also Moltmann, *Ethics of Hope*, 147.

sterile order nor the horror of sheer chaos is compatible with a God of history. Note that what plagues discourse on conservation also pertains to discourse on restoration.
7. Sustainability could be associated with God's work of conservation as long as the stability of ecosystems (and social orders) is not absolutized; the adaptability of any system is equally important for its survival.
8. There is an urgent need to relate ecumenical discourse on sustainability to ecumenical discourse on ecojustice, as these themes stand in tension with each other given divides between the Global South and the Global North (to use such uneasy terms).
9. Understanding the shift from Holocene stability to "Anthropocene" volatility as "God stirring the soup" raises more questions than it can answer. Instead, what may be needed is something like a revisiting of the Noah story, not only the role of the ark in anticipating and preparing for a volatile transition (the lifeboat metaphor is dangerous and often exclusivist, as the tragedy of the Titanic illustrates), but also a renewal of the covenant to rebuild a society no longer based upon fossil fuels and where well-being no longer assumes endless economic growth. The transition to a new dispensation will undoubtedly be volatile but the biblical notion of a covenant inspires courage, not despair, precisely amid such a transition.
10. This shift does prompt further reflection on the rubrics of *gubernatio* and *concursus*, given the anthropogenic causes of the disruption in the Earth System (noting debates on divides of race, gender, class, and caste regarding this "*anthropos*") and the ultramodern quest to displace God altogether by "playing God", indeed by becoming divine (*homo Deus* if perhaps not *theosis*).[108]

■ Bibliography

Barth, Karl. "'No!' Reply to Emil Brunner's 'Nature and Grace'." In *Natural Theology*, edited by Emil Brunner and Karl Barth, trans. P. Fränkel, 67–128. London: Geoffrey Bless/Centenary Press, 1946; Eugene: Wipf & Stock, 2002.

Bavinck, Herman. "Calvin and Common Grace." *The Princeton Theological Review* 7:3 (1909) 437–65. https://www.monergism.com/thethreshold/sdg/pdf/bavinck_commongrace.pdf.

———. "Common Grace." Translated by Raymond C. van Leeuwen. *Calvin Theological Journal* 24 (1989) 38–65. [Dutch original 1894].

Berkouwer, Gerrit C. *The Triumph of Grace in the Theology of Karl Barth*. Grand Rapids: Eerdmans, 1956.

Bonhoeffer, Dietrich. *Ethics*. Dietrich Bonhoeffer Works 5, edited by Clifford J Green. Minneapolis: Fortress, 2005.

Bratt, John D., ed. *Abraham Kuyper: A Centennial Reader*. Grand Rapids: Eerdmans, 1998.

---

108. I will be addressing the theme of God's governance in history and the problems raised by *concursus* in two other essays in association with work towards this volume.

Brunner, E. "Nature and Grace." Contribution to the Discussion with Karl Barth in *Natural Theology*, edited by Emil Brunner and Karl Barth, trans. P. Fränkel, 15–64. London: Geoffrey Bless/Centenary Press, 1946; Eugene: Wipf & Stock, 2002.

Clark, Nigel, and Bronislaw Szerszynski. *Planetary Social Thought: The Anthropocene Challenge to the Social Sciences*. Cambridge: Polity, 2021.

Coetzee, Murray H. *Die "Kritiese Stem" teen Apartheidsteologie in die Ned Geref Kerk (1905–1974: 'n Analise van die Bydraes van Ben Marais en Beyers Naudé*. Wellington: Bybelmedia, 2011.

Coetzee, Murray H., and Ernst M. Conradie. "Apartheid as Quasi-Soteriology: The Remaining Lure and Threat." *Journal of Theology for Southern Africa* 138 (2010) 112–23.

Conradie, Ernst M. "Abraham Kuyper's Legacy for Contemporary Ecotheology: Some Reflections from within the South African Context." In *Creation and Salvation: Dialogue on Abraham Kuyper's Legacy for Contemporary Ecotheology*, edited by Ernst M. Conradie, 3–146. Leiden: Brill Publishers, 2011.

———. *Hope for the Earth—Vistas on a New Century*. Eugene: Wipf & Stock, 2005.

———. "The Limits of Ecological Justice?" *Scriptura* 122 (2023) 1–12. https://doi.org/10.7833/122-1-2136

———. "The Project and Prospects of 'Redeeming Sin?': Some Core Insights and Several Unresolved Problems." *Scriptura* 119 (2020) 1–22. https://doi.org/10.7833/119-2-1689

———. *Redeeming Sin? Social Diagnostics amid Ecological Destruction*. Lanham: Lexington, 2017.

———. *Saving the Earth? The Legacy of Reformed Views on "Re-Creation."* Berlin: LIT Verlag, 2013.

———. *Secular Discourse on Sin in the Anthropocene: What's Wrong with the World?* Lanham: Lexington, 2020.

———. "Some Theological Reflections on Multi-disciplinary Discourse on the 'Anthropocene'." *Scriptura* 121 (2022) 1–23. https://doi.org/10.7833/121-1-2076

———. "Van Ruler as an Early Exponent of Christian Ecotheology?" In Arnold van Ruler, *This Earthly Life Matters: The Promise of Arnold van Ruler for Ecotheology*, edited by Ernst M. Conradie, 32–57. Eugene: Pickwick, 2023.

———. "What, Exactly, Needs to Be Sustained Amidst a Changing Climate?" In *Global Sustainability: Issues in Science and Religion*, edited by Michael Fuller et al., 25–40. Cham: Springer, 2023. https://doi.org/10.1007/978-3-031-41800-6_3

Durand, J. J. F. (Jaap). *Skepping, Mens, Voorsienigheid*. Pretoria: N.G. Kerkboekhandel, 1982.

Dutch Reformed Church. *Human Relations and the South African Scene in the Light of Scripture*. Cape Town: NG Kerk-Uitgewers, 1974.

Eliade, Mircea. *Cosmos and History: The Myth of the Eternal Return*. New York: Harper Torchbooks, 1954.

Fergusson, David. *The Providence of God: A Polyphonic Approach*. Cambridge: Cambridge University Press, 2018.

Fox, Matthew. *Original Blessing*. Santa Fé: Bear & Co., 1983.

Gilkey, Langdon. *Reaping the Whirlwind: A Christian Interpretation of History*. New York: Seabury, 1976.

Havea, Jione. "Covenant: Chose, Cover, Contest." In *reSTORYing the Pasifika Household*, edited by Upolu Lumā Vaai and Aisake Casimira, 51–62. Suva: Pacific Theological College Press, 2023.

Hessel, Dieter T., ed. *After Nature's Revolt. Eco-Justice and Theology*. Philadelphia: Fortress Press, 1992.

Institute for Contextual Theology. *The Kairos Document. Challenge to the Church*. Revised 2nd ed. Johannesburg: ICT, 1986.

Justice, Peace and Creation Team. *Alternative Globalization Addressing Peoples and Earth (AGAPE): A Background Document*. Geneva: World Council of Churches, 2005.

Kinghorn, Johann, ed. *Die NG Kerk en Apartheid*. Johannesburg: Macmillan, 1986.

Kuyper, Abraham. *De Gemeene Gratie, Deel I-III*. Kampen: JH Kok, 1930–1931.

———. *Lectures on Calvinism*. New York: Cosimo Classics, 2007 [1931].

Meadows, Donella H. et al. *The Limits to Growth: A Report for the Club of Rome's Project on the Predicament of Mankind*. New York: Universe Books, 1972.

Moltmann, Jürgen. *Ethics of Hope*. Minneapolis: Fortress, 2012.

———. *God in Creation: An Ecological Doctrine of Creation*. London: SCM, 1985.

Ned Geref Kerk, Algemene Sinode. *Ras, Volk en Nasie en Volkereverhoudinge in die Lig van die Skrif*. Kaapstad: NG Kerk-Uitgewers, 1974.

———. *Kerk en Samelewing: 'n Getuienis van die Ned Geref Kerk*. Bloemfontein: NG Sendingpers, 1986.

Neiman, Susan. *Evil in Modern Thought: An Alternative History of Philosophy*. Princeton: Princeton University Press, 2002.

Niebuhr, Reinhold. *Love and Justice: Selections form the Shorter Writings of Reinhold Niebuhr*, edited by D. B. Robertson. Louisville: John Knox Press, 1957.

Pannenberg, Wolfhart. *The Historicity of Nature: Essays on Science and Theology*, edited by Niels-Henrik Gregersen. Conshohocken: Templeton Foundation, 2008.

Presidential Climate Commission. *A Framework for a Just Transition in South Africa*. 2022. https://pccommissionflow.imgix.net/uploads/documents/A-Just-Transition-Framework-for-South-Africa-with-dedication-FSP-002.pdf.

Rasmussen, Larry L. *Earth Community Earth Ethics*. Maryknoll: Orbis, 1996.

Schreiner, Susan E. *The Theater of His Glory: Nature and the Natural Order in the Thought of John Calvin*. Grand Rapids: Baker Academic, 1991.

Sittler, Joseph. *Evocations of Grace: The Writings of Joseph Sittler on Ecology, Theology and Ethics*, edited by Peter Bakken and Steven Bouma-Prediger. Grand Rapids: Eerdmans, 2000.

Vaai, Upolu Lumā. "From Atutasi to Atulasi: Relational Theologizing and Why Pacific Islanders Think and Theologize Differently." In *Theologies from the Pacific*, edited by Jione Havea, 235–45. Cham: Palgrave Macmillan, 2021.

Vaai, Upolu Lumā, and Aisake Casimira, eds. *reSTORYing the Pasifika Household*. Suva: Pacific Theological College Press, 2023.

Van de Beek, Abraham, "Rust Is Ver te Zoeken: Over de Chaotiserende en Teken-stellende Aanwezigheid van de Geest." In *Van Utopie naar Werkelijkheid: Gebrokenheid van der Schepping*, edited by Herbert van Erkelens, 42–52. Kampen: Kok, 1989.

Van Leeuwen, Raymond C. "Herman Bavinck's 'Common Grace'." *Calvin Theological Journal* 24 (1989) 35–37.

Van Ruler, Arnold A. *Kuyper's Idee Eener Christelijke Cultuur*. Nijkerk: Callenbach, 1940.

———. *This Earthly Life Matters: The Promise of Arnold van Ruler for Ecotheology*, edited by Ernst M. Conradie. Eugene: Pickwick, 2023.

———. *Verzameld Werk Deel III: God, Schepping, Mens, Zonde*. Zoetermeer: Boekencentrum, 2009.

Welker, Michael. *Creation and Reality*. Minneapolis: Fortress, 1999.

Wolterstorff, Nicholas. *Justice in Love*. Grand Rapids: Eerdmans, 2011.

———. *Until Justice and Peace Embrace: The Kuyper Lectures for 1981 Delivered at the Free University of Amsterdam*. Grand Rapids: Eerdmans, 1983.

Evolution and Providence: A View from Aotearoa

Nicola Hoggard Creegan[1]

■ Introduction

Believers have always struggled with how God is present in the natural world and how to embrace the spirit in nature without also being idolatrous, but this is especially the case as the culture at large has become increasingly materialist and the witness to God's providence becomes more inward and individual, even a mark of special favor.

Firstly, the question has become: Does God interact with the world at all? We have moved beyond the complexities of *how* God is exercising providential care (as several authors have noted). In light of materialist science, there is no God, or God has become a deistic God. Yet an acknowledgment and discernment of the ongoing work of God is required for piety, faithful readings of the Bible, and coming to terms with the idea

---

1. Nicola Hoggard Creegan is the director of New Zealand Christians in Science / Ngā Karaitiana Kimi Matū, a chaplain at Maclaurin Chapel at the University of Auckland, chair of Te Kaunihera at St John's College, and adjunct at Flinders University.

How to cite: Creegan, NH 2024, 'Evolution and Providence: A View from Aotearoa', in EM Conradie & UL Vaai (eds.), *Making Room for the Story to Continue?*, An Earthed Faith: Telling the Story amid the "Anthropocene", vol. 4, AOSIS Books, Cape Town, pp. 89-109. https://doi.org/10.4102/aosis.2024.BK415.04

of a good God, given the suffering in the world. It is particularly needed as the "Anthropocene" progresses to enable humans to feel less fatalistic and to inspire us with the sense of God's presence even in challenging times, when the "fig tree does not blossom, and there be no fruit on the vine" (Hab 3:17). The stakes keep increasing in terms of war, social disorder, and climate disruption. The fate of the world appears to be held in the balance, increasingly vulnerable to random sudden levels of disaster and destabilization, even extinction, rather than being governed by an omnipotent, all-loving provider who would keep the margins of safety in check, or even lured by a constantly loving God, as is the case with process theology.

Secondly, in the last 150 years, the challenge to providence has come also from an evolutionary perspective. Versions of neo-Darwinism throughout the twentieth century have made increasingly difficult any understanding of God's presence in and providential governance of the natural world and its vital processes. Once, the evil of the world could be blamed on humans and a more or less literal understanding of Genesis 3. Now we must contend with sickness, death, and extinction as a part of the very web of life in which humans have arrived at the very last moment. John Haught notes:

> The main issue now as always, is that of how to reconcile evolution with the idea of divine providence. After Darwin, what does it mean to say that God "provides" or cares for the world? [...] What most perplexes theology is the Darwinian recipe for the evolution of life over the last 4 billion years.[2]

And of particular concern he marks the "brute impersonality and blindness" of the process, together with the "wide trail of loss and pain" in the evolutionary record.[3] If God is present providentially, even hidden from sight, we would expect a process a little more in tune with the nature of love and a little less random and purposeless than the evolutionary process as it has been described.

Thirdly, the "Anthropocene" adds another threat to the sense of the presence of God, this time from the future or the future/present when the very regularities of the natural world we have relied upon as the bedrock of God's sustenance might fail us. Conradie and Pearson, in this volume, make a great deal of how disrupting the "Anthropocene" is and will be, how different from life as usual in the Holocene. I argue in this essay that we need to take heed of the wisdom literature's affirmation of God's spirit/wisdom in the natural world and Paul's insistence on a cosmic Christ as a form of resistance to materialism. Also helpful is the parable of the wheat

---

2. Haught, "The Boyle Lecture," 5.

3. Haught, "The Boyle Lecture," 6, 7.

and the tares, which can dehistoricize evil, separating it from a human fall at the dawn of our species. Whatever the context in which we work, and whatever the theological tradition, the wheat can never be separated from the tares, but we can, over time, identify aspects of the good in these traditions that might make us stronger in the "Anthropocene" and leave us with more of a sense of agency and co-agency with God, even as we face increasing perils. For suffering is always worse when there is no sense of agency.

The "Anthropocene," then, beginning as it is against a materialist backdrop in the West, creates a threat to faith, similar to the prehuman opening up of evolutionary theory, but this time from the future. The natural world in the future will almost certainly become more and more unpredictable and less dependable, and yet as Pearson (and Fergusson) has argued, it was that dependability that nurtured religion and culture.[4] The future opens up in the same way that the past has in the last 150 years, revealing the strangeness of geo/political tipping points—largely hidden until they are crossed, escalating extinction, and a natural world that can no longer be taken for granted. For the first time, many people will realize that they have taken God's providential care of nature for granted, and we will have to come to terms with humans as a geo-force, binding and affecting not only the soft living stuff of the planet, but the movement of tectonic plates, ice floes, and basic chemistry as well.[5] Just as people often associate God's presence and existence with success in our personal lives, so *if* God is associated with Holocene stability, the "Anthropocene" will further threaten the traditional narrative, especially in the West where stability is more commonly experienced as normal. There is more need than ever before to revise a Western biblical theology of creation, fall, and redemption. But there is a need also to revise biology and evolutionary theory, as materialist science has been one of the drivers of the theological accommodation to materialism.

In the rest of this chapter, then, I try to offer a way of re-orienting the story of faith away from the perils of expected stability on the one hand and the Edenic fall on the other. The space between can be understood as the field of wheat and tares, and there is some comfort in knowing that tares do not signify an absence of God nor a loss of human agency.

---

4. See Clive Pearson's essay in this volume.

5. Clive Hamilton talks about the earth being "a single, dynamic, integrated system, and not a collection of ecosystems." He defines the "Anthropocene" as starting in 1945, and he stresses nonlinear dynamics and describes humans as exerting geological power. See Hamilton, *Defiant Earth*, 4–11.

■ Materialism: The Backdrop

Twentieth-century Western faith accommodated a materialist worldview in many ways, replacing more holistic and interdependent understandings of the world and of human and creaturely agents. In this materialist scenario, the individual human is the only consciousness that matters, and the salvation drama is all about humans and God. The fall looms large in this story, and Christ's sacrifice or substitutionary death then provides the cure. The drama of the natural world, the comfort of God's presence in nature, and questions about the status of humans relative to animals are all obscured. Evolutionary theory is often rejected because it opens up an arena of suffering that was previously unexplored and unacknowledged and because, if accepted, it would invalidate the fall as a tidy explanation for all suffering and death.

This literal storyline continues and persists in many of our Christian communities, as Van den Brink has pointed out.[6] Nevertheless, even within this story, humans depend on nature, praise God for its beauty and provision, and often blame God when natural disasters happen. Disasters make us question God, but we quickly return to our half-hearted sense of God's presence. Materialism obscures but does not obliterate the lived expectation of God's providence. Evolution could always be ignored or denied, but the "Anthropocene" will be and is in our faces. It cannot easily be ignored or accommodated.

In this essay, I examine the tensions and polarities existing in current theology and church life, under pressure from the "Anthropocene," especially in Aotearoa/New Zealand. And I suggest ways in which a new theological emphasis, new scriptural sources and new biology can reorient us to a place of resilience in the difficult times ahead. And in Aotearoa, we look forward to a retelling of the faith story that incorporates both Western and Indigenous modalities.

■ Scripture

Across traditions, the story that most pertained to a theology of creation and providence was Genesis 1–3. This is the first and most iconic creation story. It can be read in many different ways, and it encompasses creation, the first indications of God's providence, and the "fall," the foil against which the rest of sacred history is almost always read. Genesis 1 and 2 can be understood as creation by fiat and/or as the creation of powers and agents that are

---

6. See Gijsbert van den Brink's essay in this volume.

themselves decisive and generative, already with hints of providence.[7] Genesis 1 and 2 can be understood as creation out of nothing or instead as the creation of order out of chaos, as an active personal God always present, or indeed as the God who is now resting and relatively absent.

Genesis 3 is the kernel of systematic theology and of popular piety. It was the place where religious people became odd to the world. A paradise, a tree, an arbitrary command, the seeming innocence and curiosity of the transgression. The vast ongoing effects are out of all proportion to the offence. Theology of fallenness had its own grammar in the way in which people thought about the world. All evil was explained by some sort of quasihistorical act, even for those who didn't believe it really could have happened. Or prehistory was so far away that it could be vague, and origins could be evaded. The fall was so big that it could be real and not real, historical and ahistorical, a symbol and yet rooted in the past so tenaciously that it became an alternative history, easily but not always expressed as creationism or intelligent design. But the fall (and creation) story also preserved anti-ecological perspectives: humans were fundamentally different from animals. And it obscured the ongoing work of God in creation, a work that is underscored by evolution but not a work of six days, however prolonged.

In an earlier book, I argued that this story, even in its mythical form, is not adequate to account for the eons of a changing planet and an evolving story of life, and that this story is one that invites huge swathes of the church to be stuck in a literal quagmire.[8] In that book, I extended the meaning of the parable of the wheat and the tares and projected it back into nature, which we now know was full of disease, decay, death, and extinction—along with beauty and perfection—long before humans appeared. The parable allows us to think backwards, accounting for the interactions of good and evil in the contingent time-bound experience of matter. I will extend that argument here, showing that a wheat-and-tares world can give meaning to the phenomenon of increasing evil alongside goodness, and it can reassure us that a divine presence is still with us and will bring all things to a conclusion. The parable can also encourage us to look at the fine print, discerning that both good and evil, perfection and misalignment are always present.[9]

---

7. N.T. Wright for instance, talks about "mutual relations within the very being of the one God: a to-and-fro, a give-and-take, a command-and-obey, a sense of love poured out and love received." See Wright, "Jesus and the Identity of God."

8. For a longer discussion, see Hoggard Creegan, *Animal Suffering*, chapter 6.

9. Hoggard Creegan, *Animal Suffering*, 89.

Indeed, we know that the natural world, from the beginning, has exhibited deep flaws, if not of structure, at least in terms of integration. And yet nature also reveals extraordinarily dense levels of design and order and felicity and nomic regularity. Elaine Wainwright, in exegeting the Gospel according to Matthew, says of the parables that "they give us whispers of divine presence with us/with Earth." She continues, "rather than being reduced to univocal meanings related to human behavior, the parables are able to function in ways that draw readers/hearers into greater attentiveness to Earth processes."[10]

The parable of the wheat and the tares makes sense of why it is always so hard to separate out the good from the bad, how often they are infuriatingly close. As humans, we are always trying for purity, for completion, but a part of the mystery of being in nature is that everything takes time, and in time, we can never have the perfection of relationship or direction, even without the factor of the possibility of a malignant presence in all human/creaturely affairs. The wheat and the tares is a nonhistorical way of understanding the ongoing presence of evil and the difficulty of really disentangling it from goodness. Also explained with the parable is the way in which good and evil can both grow in history; the stakes keep getting higher and higher, something we see happening in the "Anthropocene." Stephen Pinker can still look at the world and decide that violence overall is less severe than it was, but we know that something is wrong because we are close to one little human error causing annihilation.[11] And increasing levels of technology make the possibilities of ruptures and ends to tyrannical regimes much less likely than they once were, as is seen in places like North Korea, China, and Russia. The parable of the wheat and the tares allows for the growth of the tares, but the sower is still there, concerned and present in the field. The presence of evil does not signify the absence of the sower, only the way in which grace and tragedy co-exist in space and time, and that God is not fully endorsing all that happens all the time. Only at the end of time.

Ernst Conradie speaks of the need for adaption—and not just sustainability—in the "Anthropocene," that we cannot return to a Holocene stability.[12] The parable of the wheat and the tares, applied to nature, as Wainwright has suggested, is also helpful in understanding that wheat and tares will continue to be woven together in this new epoch and that the precipitous growth of tares should not discourage to the point of despair. This parable will also be helpful in explaining and understanding the

---

10. Wainwright, "Hear then the Parable," 140–41.

11. Pinker, *Better Angels of Our Nature*.

12. See Ernst Conradie's essay in this volume.

"Anthropocene," along with the wisdom literature and other stories of God's presence in times of despair, as being more appropriate metaphors and narratives for both providence and an explanation for ongoing evil in and through the "Anthropocene."

The Book of Wisdom also speaks to the lively, attentive, ongoing, and anticipatory presence of God in nature. Of Wisdom it is said:

> There is in her a spirit that is intelligent, holy, unique, manifold, subtle, mobile, clear, unpolluted, distinct, invulnerable, loving the good, keen, irresistible, beneficent, humane, steadfast, sure, free from anxiety, all-powerful, overseeing all, and penetrating through all spirits that are intelligent, pure, and altogether subtle.
>
> For wisdom is more mobile than any motion; because of her pureness she pervades and penetrates all things. For she is a breath of the power of God, and a pure emanation of the glory of the Almighty; [...].
>
> Although she is but one, she can do all things, and while remaining in herself, she renews all things; in every generation she passes into holy souls and makes them friends of God, and prophets; She reaches mightily from one end of the earth to the other, and she orders all things well.[13]

This is a blueprint description for an entangled world, one in which the Spirit of God is present in souls and in other life forms. The anthropologist Tim Ingold talks about the way in which intelligence and wisdom are not an inwardness that is contained within the rigid contours of a body; rather, they are expansive and interactive and enlivening, like the breath of God or the Wisdom of God.[14]

■ Materialist Science

It is important to understand, though, why these texts which speak so clearly of God's close work within nature register very little in the modern context. Materialist science makes the wisdom literature look like poetry, and it has pushed the rich history of interpretation of Genesis in one direction, that of deism, and away from the subtle God of creation and providence. I argue that one of the problems for providence in this age—of reductionism, materialism, and rationalism—is that creation and providence have been conflated. All activity of God is understood as top-down, human-like, the kind of activity that might have been present in creation. It is creation out of nothing, rather than through the medium of other causes and minds. When the idea of a God who acts like this is rejected as implausible, so is the whole of idea of God.

---

13. Wisdom 7:22–25, 27; 8:1 (New Revised Standard Version).

14. Ingold, "The Major and the Minor."

This is true also of evolutionary interpretation. Twentieth-century interpretations of Darwin posited a directionless, purposeless evolutionary process—a clever algorithm, going nowhere in particular, reconcilable perhaps with deism, but not with any understanding of providence, or at least any kind of providence that could be seen.[15] In spite of all of this, God was never completely evacuated from nature, and faithful people did thank God for the goodness of creation, for its bounty, for its beauty and its provision. But if we add in now the idea of Earth System disruption in the "Anthropocene" … human societies and the churches are reaching a tipping point where the natural reassurances of God's ongoing presence and work are fast being undermined.

The second half of the twentieth century in the West was particularly dire in its reductionism and materialism. Yet if providence is true, God is intimately at work in the whole process in some way, and surely we should be able to see some of this. Elizabeth Johnson says of God's actions in history that they cannot be understood to be completely hidden and obscure: "When received in a faith context, historical research can indeed strengthen as well as challenge faith, for divine presence and action in the world are not so intangible as to leave no discernible historical traces."[16]

The same can be said for God's action in creation.[17] Instead, natural selection had become a mantra for the twentieth century in Western science. Famously, the (United States) National Association of Biology Teachers (NABT) stated quite plainly, mid-century, that "[t]he diversity of life on earth is the outcome of evolution: an unsupervised, impersonal, unpredictable and natural process of temporal descent with genetic modification that is affected by natural selection, chance, historical contingencies and changing environments."[18]

This statement was both hugely influential way beyond the confines of North America, and in turn reflected the zeitgeist of late twentieth-century biology, leaving faith no alternative apart from fideism. It was reluctantly changed in 1997 under pressure from the philosopher Alvin Plantinga, among others. But the late twentieth century was not a place hospitable to

---

15. Hoggard Creegan, "Re-engaging Theology and Evolutionary Biology."

16. Johnson, "The Word Was Made Flesh," 149.

17. Nevertheless, Swinburne and others have argued that God must be partly hidden in order for there to be genuine freedom in good and evil. If God were so compelling we could not ignore God, this would be like holding a gun to our actions. We would have no choice but to do the good. See Swinburne, *The Existence of God* (revised).

18. The controversy over this statement is discussed in a *Christian Century* in 1997, and by the American Scientific Affiliation (ASA), in 1997, https://www.asa3.org/ASA/education/origins/nabt.htm [last accessed October 10, 2024].

intimations of God in nature or the providential actions of God, whether it be the "Anthropocene" or the Holocene.

Evolutionary biology was often opposed to hints of God in nature. Natural selection alone appeared to eliminate all need for a personal God, one in which God's providence could be observed. Darwin at least played with this idea. He knew that it broke the connection between Creator and creation. One could dissent; process theology filled this niche to some extent, but it always appeared to be over-confining of God's powers; panpsychism, now more seriously considered, comes a lot closer to being a satisfactory (vitalist) metaphysics from a Western ecological perspective.[19]

■ Tradition and the Tensions in Theology

The providence of God, unlike the bare facts of creation, is understood and experienced locally and personally. Hence, tradition is relevant. I should admit at the start to my own tradition being stratified or liquid, lived on the boundaries of various tensions, and underlining for me the conflicts the church embodies, but always within the broad stream of thinking that is the border between science and theology. I am an Anglican, but my early influence and immediate family were Roman Catholic, lived in Aotearoa / New Zealand, a bicultural Treaty-based, largely secular society (apart from Mātauranga Māori, or Indigenous ways of knowledge, which are now very center stage in this society) with large Polynesian and Asian populations, very much at the ends of the earth. My theological training was Reformed within a Methodist university in the United States of America (USA), and undertaken in the maelstrom of race and feminism in the late eighties in the USA. I taught for 13 years in an evangelical theological school, attempting to bridge (without much success) the theological chasm between evangelicalism and a more liberal faith. Although I am an Anglican, few Anglicans here are interested in the science/faith boundary today—another boundary.

Anyone trained in theology will recognize that there are almost unbearable tensions in Christianity today, such that commitment to one Christian church will pit one against members of another, so great is the intellectual and spiritual gulf between them; certainly, also, there are strengths and flaws, wheat and tares in every tradition. For instance, the Roman Catholicism of my childhood accorded me two visions. On the one hand, there was the often-reinforced doctrine of eternal damnation and mortal sin and an insistence that only the elect and those humans who may be said to be baptized, even if only by desire, were saved. Humans were

---

19. See, for instance, Nagel, *Mind and Cosmos*, and Leidenhag, *Minding Creation*.

saved and not animals. On the other hand, there was the experience of glory in high holy days, the constant reminders in statues and incense and chant and Eucharist that this glory was offered to human beings, but they also seemed to emanate out of matter and creation, to be not just the domain of humans. This latter collection of images and foci are the ones that have endured, that gave freedom and a sense of the expansiveness of God's love. In continuity with this was my early exposure to Teilhard de Chardin and his synthesis, which has never left me, of the evolutionary process being wrapped up into an omega point in Christ. Science and faith together always transcended either enterprise alone.

Between the early intense childhood faith and the late twentieth century, I learned a lot of mathematics, biology, and theology and had begun to put them together again. In my doctoral dissertation, I interacted with both Jonathan Edwards and Friedrich Schleiermacher, both Protestant but fathers of quite different traditions, and both very familiar with the science of their day. Edwards's was a vision of love embedded in nature in its symmetries, surprises, and beauties (at least in "The Nature of True Virtue," if not in some of his other writings).[20] At the same time, I had to reject the Calvinism and vivid images of hell and associated fear which were evident in *Sinners in the Hands of an Angry God*. Edwards was a conflicted character. I was attracted also to Schleiermacher's generous orthodoxy, his polymathic knowledge of both science and the arts, the rooting of all piety in the human consciousness as it interacts with the whole, and with the sense of dependence that the whole engenders.[21] I could affirm his passing recognition of continuity in consciousness with infants and animals, and his universalism in Christ consciousness, like Teilhard de Chardin (but so unlike Edwards). All of these were woven eventually into my ecotheological consciousness. And it was human consciousness and its connections to matter which would persist.

But feminism and race emphases were also important (James Cone, Catherine Keller, Elizabeth Schussler Fiorenza, Rosemary Radford Ruether, Kwok Pui-Lan, and so on). This engagement with science and theology in light of liberation theologies opened up the world of ecotheology; surely nature also needed liberating, or at least attending to. I understood that Lynn White Jr. was partially right, that Christianity in late modernity, and in thrall to Barthianism or fundamentalism, did train people who were oblivious to nature, afraid of its powers as idolatry, and uncaring of its charms.[22]

---

20. Jonathan Edwards argues that patterns in nature are a lower form of love. See Edwards, "The Nature of True Virtue," 521.

21. Schleiermacher, *The Christian Faith*, 12ff.

22. White, "The Historical Roots of Our Ecological Crisis."

Similarly, in science, people were trained to see nothing but natural selection even if they secretly believed that God's creative power was out there somewhere. In the process of this reflection, I found my own inner reaction to nature and to animals (and, indeed, later to plants) being transformed, insofar as anyone who is the product of late modernity can be transformed. Something was going on in nature that was powerful and alive and not just the throw of the dice. Animals mattered. Perhaps even plants.[23] To the extent that persons can feel the presence of God, it is, as often as not, through the medium of the natural world and its vitalities. This is most evident and is reinforced in Aotearoa because Mātauranga Māori is so centered around *mauri*/life force and *whenua*/land.

I have increasingly wanted to affirm and magnify everything in the tradition which spoke to the weaving of love into the fabric of creation, its interconnections, its vitality, and its almost infinite depth and surprises. Humans were a part of creation, sharing its history and its DNA and the constraints of matter and life form. Humans are also different, speaking animals with a desire to know and speak with God, and with the audacious idea that life extends beyond and apart from this life. At this point in history, as we enter a time of increased vulnerability and uncertainty, biology is again beginning to embrace the mystery and depth and purposefulness of the natural world, as I summarize in the last section of this essay. Yes, we play with the idea of a Gaia, but Gaia then lets us see the wisdom of the wisdom literature and of a more holistic understanding of Genesis. The history of life on earth is not passive until it comes to humanity, as suggested for instance by Clive Hamilton in *Defiant Earth*;[24] rather, it is the outworking of consciousness and love and agency at every level of the evolutionary tree.

Thus, Christianity, like our story of origins, often presents two visions of God and humanity, of sin and redemption in the same era, and sometimes in the same tradition or the same denomination. As we consider the "Anthropocene," it is possible to choose from these strands the ones that are most plausible and have endured in an ecological and nature-attentive theology; those that allow the widest possible scope for God's involvement with the experiment of matter and consciousness in matter; those that see continuities between humans and other life forms (even while affirming human uniqueness), that embrace the widest possible vision of salvation, that allow that God's energies, God's glories are touching the natural world and evident within its powers and can be glimpsed even if not understood; theologies that see *Wairua*/Spirit as a wider domain of reality than matter

---

23. See for instance, Gagliano, *Thus Spoke the Plant*.

24. Hamilton, *Defiant Earth*, 6.

alone; theologies that understand humans as Janus figures, pointed downwards toward the earth and upwards to communicate with god and the angels and the communion of saints. In the conflicting traditions I have been a part of is the idea that human consciousness bears a particular capacity for God consciousness and a particular responsibility for the care and healing of creation in partnership with God.

■ The "Anthropocene" in Aotearoa / New Zealand

As well as tradition, one must speak also of context. The "Anthropocene" in Aotearoa / New Zealand has a number of interesting aspects. Early European settlers (or invaders) to New Zealand intentionally introduced mustelids (in order to control an earlier influx of introduced rabbits) to a country without any native mammals apart from bats. This ended up being a disruption in itself, as New Zealand's unique bird population largely nests in burrows in the ground or accessible in trees—easy game for a new species with precocious fertility like stoats. The mustelids added to the problems of rats which came in two waves, first the Polynesian rat with the Māori (which did impact the flora and fauna but not as greatly as subsequent rodent waves) and then the ship and Norwegian rats with the settlers' boats.[25]

The official policy now is a predator-free New Zealand, with intense killing campaigns, using bait and traps in many protected areas, but also in domestic neighborhoods.[26] In other words, prodigious resources are going into returning New Zealand to its prehuman pristine state, or at least its pre-European state. Yet, if we take the longer view, evolution has never been static. A part of our understanding of God's presence in a rapidly changing climate must be to understand how we can creatively leverage the death of an earlier era in the co-creation of a new. I give several examples below.

Firstly, I am a part of the Christian environmental group A Rocha and our flagship project in Raglan is in the business of the intensive killing of invasive species.[27] This is to save not only the bush but also, importantly, the key species of the migrating mutton bird on the mountain, and to make the mountain safe for other endangered species like the kaka. But is this

---

25. See https://predatorfreenz.org/toolkits/know-your-target-predators/rat/ [last accessed October 10, 2024].

26. See https://predatorfreenz.org/toolkits/know-your-target-predators/rat/ [last accessed October 10, 2024].

27. See https://aranz.org.nz and https://karioiproject.co.nz [last accessed October 10, 2024].

killing something that is a part of God's providential plan? How do we understand the presumed divine favoring of one species over another when the aim is to bring about a previous state of equilibrium of a world we know has gone through many eras and eons of disruption in the past? If we give up the recent past as the ideal state, does that alter our policies or killing program? How can a wheat-and-tares framework reconcile us to the need to do violence to some species so that others might live?

Secondly, some of the offshore islands of New Zealand are predator-free but others, like Ponui, have limited numbers of predators, in particular small populations of feral cats and rats which seem to keep each other in check. The policy is to leave them be, which is an example of adaptation rather than blindly proceeding to eradication.

A third pressing "Anthropocene"-related problem also looms for all Pacific islanders—the rise of sea level. Many New Zealanders are within an easy drive to the coast or on coastal plains that are less than 10 meters above sea level. Policies have already been made for the fair and programmed withdrawal from the sea, but these are also already producing huge concern among homeowners who see their property prices drop drastically long before a real problem emerges. Few of these homeowners would believe that God will save them, but few also would acknowledge the urgency of a situation that happens very slowly.

Also important are the cultural disruptions of the "Anthropocene" in New Zealand, beginning with the colonial impact and the ongoing clash of cultures in a country that is experiencing a renaissance of Indigenous language, culture, and spirituality, together with a right-wing reaction to any special treatment or collectivist policies. Numerous questions are active at this boundary. The current government is a case in point. A coalition of two minor parties and a more mainstream conservative party have now replaced a Green/Labour coalition that brought in much progressive legislation. At the extreme right, the Association of Consumers and Taxpayers (ACT) party leader talks about the rights of the individual and their threat under the current collective politics and mindset. Collective policies, including those of the original Treaty of Waitangi (signed in 1840 between a fledgling British governance team and most Māori chiefs, resurrected as a living document in the 1970s), have no place in a modern democracy, they argue. That is huge in a country where the rekindling of the Treaty and the processing of land claims through the courts has been central to life in New Zealand for decades and has arguably produced helpful social cohesion. Along with the individualist policies is a disregard for the environment and climate change, while the Indigenous stance is one that has seen the Whanganui River, for instance, achieve personhood status in law. A culture with a strong sense of nature's forces and spirits as a part of the *whanau*/community and with strong ties to *whenua*/land/ancestors is interacting with cultures that are at the tail end of the Western Enlightenment. This culture (in its more peaceful guises) resonates strongly with our emerging sense within the Western tradition of theology and biology of agency and apprehension existing at every level of the life project. The overall effect of this melding of cultures is that New Zealanders can more easily embrace the importance of the agencies of moa, rats, stoats and birds—and also ancestors—in the collective story of our *whenua*/land.

The tension inherent in New Zealand society is a lens through which we can view the wider society and ecclesial landscape. Although in Aotearoa we live in a world with volcanoes and earthquakes, which gives us a theoretical knowledge of how unstable our landscape is, nevertheless, like others, we build cities as though stable times will continue forever. We build houses and highways on earthquake fault lines and houses perched on the edges of steep cliffs or on plains looking out over a sea that is encroaching. Our obsession with predator control, important as it is, sometimes substitutes for more in-depth thinking about the changes ahead and the challenges, as though getting back to a time without rats and stoats will solve our problems. The full impact of the change that is ahead is now only beginning to be a part of the discussion in our civil society or in our theology.

■ Theological Tensions

My own history and theological engagement have navigated the polarities existing in Christian theology and the Christian church, most of which impact on the theology of providence.

Western theology, including that in Aotearoa, strongly mirrors society's materialism, especially, ironically, in some more conservative churches. The primal story remains the fall narrative, with the mistakes of one human being (or two) responsible for all evil in the world, regardless of the clash this scenario sets up with science. Theodicy is not engaged in with any reference to the full sweep of history but rather in reference to this primal story. There is little motivation, then, to listen for God's voice in the deep sorrow of the world. Or to look for God's providential and mysterious presence in the natural world.

At one end of this theological extreme, all actions of God are understood as being direct, as in creation, and increasing evidence of God's absence may result in despair or, alternatively, happiness that the end is upon us. Extreme individualism, including the sense that the individual will succeed or fail to end up in heaven or hell, is more associated with understanding

Christ's death as a sacrifice or a substitution made in history and appropriated through sacrament or faith for the individual person. Because these theologies are so based on the individual's appropriation of faith, they do not move into the sphere of nature, consciousness, or humans as a species. And they do not sit easily with the idea of God's providence, except as God's offer of salvation and sanctification to the individual.

Many adherents to these forms of Christianity are intensely propositional; they are anti-science or at least anti-evolution, although increasing numbers are embracing intelligent design, which in most of its manifestations does incorporate God's direct actions at points along the way of an otherwise impersonal process.

Any kind of materialist Christianity will find a rapprochement with Indigenous thought impossible, and tensions will remain as forms of syncretic Christianity are rejected.

The individual *can* have faith against a backdrop of a materialist view of nature as essentially mechanical and without a point of view. But the more universal consciousness-related family of theologies incorporates an understanding of prehension, or consciousness or concrescence, as always imbued with and in response to providence and the more subtle call of God.

In these more continuous universal and incarnational theologies, as modeled by Schleiermacher, Teilhard de Chardin, Tanner, Coakley, Rohr, and so on, salvation is understood in the widest possible way and is related to the spread of God's consciousness, love, or inclusion in the corporate life of Christ and the Wisdom of God in creation. These latter models would incorporate a more universal final application while acknowledging that the wheat and tares continue to grow in creation. This family of theologies is consistent with an understanding of providence as divine action that is more subtle and interacts with and allows the forces and agents that constitute creation to continue. Such an understanding of theology encourages hope in a future of divine–human partnership, rather than always harking back to creation modalities or salvation by substitution.

Despite newer developments in theology and biology, there is no doubt which of these strands is still predominant in Western-influenced countries. We reach this period of the "Anthropocene" at the end of a 500-year period of Enlightenment, with increasing materialism, increasing rationalism (or left-brain thinking), and increasing reductionism and mechanization. The individual is paramount and is often thought to be the only mind-like substance on the planet. Life is the result of an impersonal algorithm without any sense of purpose. The community and the common good are all abstract terms that bear little weight when compared to the individual.

The clash of these different ways of viewing the world is particularly evident in Aotearoa / New Zealand, where a Western-influenced *pākehā* culture meets an Indigenous Māori one which is always resisting such materialism and reductionism.

At the other end of the theological spectrum, providence becomes coherent as the subtle interactions of God with the generative powers, agents, and decisions of all creation. In this paradigm, it still becomes meaningful to understand our actions as cooperative with God in bringing peace, adaptability, new ways of surviving, and new creative and less fatalistic forms of social organization and resistance. An understanding of God's actions through the parable of the wheat and tares does not insist that God is endorsing all that happens but instead prepares us for the sober realities of both wheat and tares growing in history.

Providence makes the most sense within a theology that affirms that God joined us in this experiment of matter program, and that matter is a form of consciousness that allows stuff to happen in time. Believers, like Indigenous people, must assume a much wider and deeper cosmology than matter alone. Salvation is achieved as Christ's consciousness invades and possesses us as people, including the grammar of the Sermon on the Mount, the rubric of forgiveness, the achievement of social justice in its widest sense, and the embrace of weakness over strength. Nature is a home for *wairua*/Spirit without meaning idolatry; rather, the creator God has taken up residence in nature, can be discerned within its processes, and is deeply present to all natural cycles and patterns of growth, in the forces (sometimes personified and sometimes not) we understand and in those we do not. But God is not contained in this experiment with matter but rather rules within and beyond matter. God's providential actions importantly do not conflict with the creature's actions.[28] Thus dreams, intuitions, and signs can be present in nature without God intervening heavy-handedly. God's spirit and the overall trajectory of nature are of one accord.

Humans, more than any other created agent, can choose to act against the demands of the Spirit or can be collected up into a conscious or an unconscious rebellion. Knowledge of the Christ is not a magic that works salvation by itself but rather helps the individual choose to join and remain within the arc of salvation, defined in nature, in consciousness, and in human and creaturely society. Theology in this more expansive dimension has some hope of making meaning even within the "Anthropocene," even as defined by Earth System disruptions and tipping points.

---

28. Kathryn Tanner talks about the way in which God's actions and human actions are non-competitive. See Tanner, *Christ the Key*, 90.

■ Changes in Biology That Make a Difference

But it is not just theology that deserves attention; so too does biology. There is now a great deal going on in biology that would resist materialism, that would push back against reductionism. In biology, we have a rapidly changing paradigm of evolution, and this has immense significance for a theology of providence. What was once profoundly and deliberately hidden—the hand of God in nature—is now open to the light of discernment, as is an almost limitless depth in biology. Changes in biology, which I will summarize briefly below, do not at all, of course, demand a theist interpretation, but they do present a puzzling and complex scenario that is no longer an easy fit with either deism, atheism, or even intelligent design; rather, there appears to be a form of inherent agency, openness, attentiveness, and anticipation—even altruism—within the process of life that fits instead with the idea of wisdom. Certainly, the contemporary understanding of evolution is multifaceted, as though a whole new domain has been opened to scrutiny and reveals not only the dense, layered complexity of life but also its ongoing progress. There is in some ways no easy demarcation between creation and providence if we examine this from a theological perspective.

One of the major contributors to the contemporary awareness made early in the twenty-first century was in the work of Simon Conway Morris, the world's expert on evolutionary convergence, or the phenomenon by which the same evolutionary solutions keep occurring again and again, something we would not expect in a completely random evolutionary world. Evolution seems to know where it is heading.

He has said famously, "The heart of the problem [...] is to explain how it might be that we, a product of evolution, possess an overwhelming sense of purpose and moral identity yet arose by processes that were seemingly without meaning. If, however, we can begin to demonstrate that organic evolution contains deeper structures and potentialities, if not inevitabilities, then perhaps we can begin to move away from the dreary materialism of much current thinking with its agenda of a world now open to limitless manipulation."[29]

Conway Morris is a Christian, but most of the other major players in the paradigm shift are not. Some anthropologists have also become uneasy with the reductionist Darwinism, dealing as they do with the evolution of conscious agents in history. They have formulated what they call the "Extended Evolutionary Synthesis," and the idea of evolution within a niche. Agustín Fuentes, for instance, says:

---

29. Conway Morris, *Life's Solution* 2. But see also his *Six Myths*, a dense compendium of all the ways we have been misled by the evolutionary theory of the twentieth century.

> [I]f we are to take human cultural processes, in all their complexity, as part of our evolutionary approaches, we cannot treat them as a social, material, historical, and perceptual veneer laid over a basal set of physiological capabilities. In human evolution, the biological and social cannot be seen as distinct entities. [...] A contemporary evolutionary approach has to treat what humans do and experience as a complex system that has specific histories, has inherited ecologies and institutions, and includes a myriad of categories of action and perception as they relate to the interactions between individuals, groups, and the communities in which they exist.[30]

Philip Ball has written a series of books looking at the role of physics and chemistry in evolution. Again and again, an evolutionary solution is the result of a best fit in terms of chemistry, rather than the outcome of random attempts at different possibilities. Nature will choose, he says, "to create at least some complex forms not by laborious piece-by-piece construction but by harnessing some of the organizational and pattern-forming phenomena we see in the non-living world."[31] Stripes and forms and patterns on animals and in cell division all follow similar deep physics and chemistry.[32]

James Shapiro, in a ground-changing essay on evolution, recently said: "A lot has changed since 1859. We now know that Darwin's 'gradualist' view of evolution, exclusively driven by natural selection, is no longer compatible with contemporary science. [...] It's not just that random mutations are one of many evolutionary processes that produce new species; they have nothing to do with the major evolutionary transformations of macroevolution. Species do not emerge from an accumulation of random genetic changes."[33]

He goes on instead to talk about the importance of symbiogenesis, horizontal deoxyribonucleic acid (DNA) insertions and transposition, and the way in which DNA is built up from the insertion of fungi, bacteria and viruses. Graeme Finlay, similarly, talks about the placenta and the three independent insertions of transposable viral elements that were required to make this most essential of mammalian organs.[34]

Shapiro adds: "My own belief is that the reason for this wilful neglect lies in the basic philosophical foundations of mainstream thinking about evolution, which requires a purely physical explanation for all evolutionary processes."[35]

---

30. Fuentes "The Extended Evolutionary Synthesis," 16–17.

31. Ball, *Shapes*, 17.

32. Ball, *Shapes*, 152–92.

33. Shapiro, "Evolution without Accidents."

34. Finlay, "Amazing Placenta," 306.

35. Shapiro, "Evolution without Accidents."

Even from this brief review, we can see that the world of evolution is very far from the blind impersonal process it was thought to be in the mid-twentieth century. Evolution has shown itself to be hyper-adaptable, coming up not just with minor variations but with jumps and sudden changes as the circumstances allow and prescribe. Van den Brink talks about the need for an understanding of God in terms of interruption, transformation, and liberation.[36] And about how the God of providence is also the Father of Christ the incarnate one. It is interesting, then, that the story which was at the heart of the materialist narrative for so long is now radically changed, if only we could get our head around it and believe it. There is now more reasonableness behind the option for connectedness, for a Wisdom-inspired theology, than there ever was before.

■ Conclusion

To summarize the arguments above, I have argued that the "Anthropocene" opens up the future to theological scrutiny around providence, in the same way that the deep past has with evolutionary history. I adapt the parable of the wheat and the tares to acknowledge the growth of increasing disruption in the future, as the "Anthropocene" changes human and creaturely life on this planet. I acknowledge also the strands of Scripture from the wisdom literature to Paul's cosmic Christ where the deep penetration of all matter by mind and spirit is insisted upon. We are not left alone with the disruptions and surprises the "Anthropocene" will bring.

The parable also encourages the kind of deep scrutiny of the material which might allow us to see that both tares and wheat are present. The changes in biological sciences around evolution have allowed this; we can now see that although there are random elements, these are constrained and the whole process is much deeper and more creative and more cooperative than was once assumed.

But I apply the parable also to the mix of helpful and unhelpful aspects of all our traditions and contexts. I argue that the individualistic gospel, which notices only humans as objects of God's concern, is less of a gospel for the "Anthropocene" than the traditions that acknowledge the deep wisdom always in nature, the primacy of Spirit in creation, and the importance of Incarnation for all creatures and not just an elect few.

I argue in the end for a holistic vision of interconnection and the primacy of consciousness and a sense of mind in nature as most robust when encountering the "Anthropocene." I argue that these holistic visions are emerging slowly from our traditions but that biology is also changing

---

36. See Van den Brink's essay in this volume.

greatly and that the changes amount to a paradigm shift when compared to the materialist science of a generation ago. In Aotearoa, as we bump up against a strong Indigenous Mātauranga Māori (Māori ways of knowing), there is an advantage in living with and alongside a culture that has always understood the deep interconnectedness of matter and spirit.

This means that although we will continue to anguish over God's presence or absence in times of devastation and climate change, nevertheless theology, traditions, and biology also give us resources for the journey and for resilience in less stable times ahead. Scripture has always been realistic about the times of hardship, and the cross has given us a model of sacrificial redemption, but it is human nature to always want good things for ourselves and our children. The grand sweep of evolutionary and geological history helps us to put things in perspective: Holocene stability has been unusual. Life has always found a way to survive. More than that, though, biology has allowed us to glimpse the infinite depth of nature's provision, adaptability, and evolvability for the first time since Darwin. And for the first time, there are theological and scientific resources to allow us to renew and reframe the broad luminous vision first presented prophetically by Teilhard de Chardin in the early twentieth century.

■ Bibliography

Ball, Philip. *Shapes, Nature's Patterns: A Tapestry in Three Parts*. Oxford: Oxford University Press, 2009.

Conway Morris, Simon. *From Extraterrestrials to Animal Minds: Six Myths of Evolution*. West Conshohocken: Templeton Press, 2022.

———. *Life's Solution*. Cambridge: Cambridge University Press, 2005.

Edwards Jonathan. "The Nature of True Virtue." In *The Works of Jonathan Edwards, Vol. 8, Ethical Writings*, edited by Paul Ramsey. New Haven: Yale University Press, 1989.

———. "Sinners in the Hands of an Angry God." DigitalCommons@University of Nebraska, 1741. http://digitalcommons.unl.edu/cgi/viewcontent.cgi?article=1053&context=etas.

Finlay, Graeme. "The Amazing Placenta: Evolution and Lifeline to Humanness." *Zygon* 55:2 (2020) 306–26. https://doi.org/10.1111/zygo.12590

Fuentes, Agustín. "The Extended Evolutionary Synthesis, Ethnography and the Human Niche: Toward an Integrated Anthropology." *Current Anthropology* 57:513 (2016) S13–26. https://doi.org/10.1086/685684

Gagliano, Monica. *Thus Spoke the Plant*. London: Penguin, 2018.

Hamilton, Clive. *Defiant Earth: The Fate of Humans in the Anthropocene*. Malden: Polity Press, 2017. Kindle Edition.

Haught, John F. "The Boyle Lecture 2003: Darwin, Design and the Promise of Nature." *Science and Christian Belief* 17 (2005) 5–20.

Hoggard Creegan, Nicola. *Animal Suffering and the Problem of Evil*. Oxford: Oxford University Press, 2013.

———. "Re-Engaging Theology and Evolutionary Biology." In *The Evolution of Wisdom: Major and Minor Keys*, edited by Agustín Fuentes and Celia Deane-Drummond. Pressbooks, 2018. https://pressbooks.pub/ctshf/chapter/re-engaging-theology/.

Ingold, Tim. "The Major and the Minor." In *The Evolution of Wisdom: Major and Minor Keys*, edited by Agustín Fuentes and Celia Deane-Drummond. Pressbooks, 2018. https://pressbooks.pub/ctshf/chapter/evolution-in-the-minor-key/.

Johnson, Elizabeth. "The Word Was Made Flesh and Dwelt among Us: Jesus Research and Christian Faith." In *Jesus: A Colloquium in the Holy Land*, edited by Doris Donnelly and J.D.G. Dunn, 149. New York: Continuum, 2001.

Leidenhag, Joanna. *Minding Creation: Theological Panpsychism and the Doctrine of Creation*. London: Bloomsbury, 2021.

Nagel, Thomas. *Mind and Cosmos: Why the Materialist, New-Darwinian Conception of Nature is Almost Certainly False*. Oxford: Oxford University Press, 2012.

Pinker, Stephen. *Better Angels of Our Nature: Why Violence Has Declined*. New York: Viking, 2011.

Schleiermacher, Friedrich. *The Christian Faith*, edited by Hugh R. Mackintosh and James S. Stewart. Philadelphia, PA: Fortress, T&T Clark, 1976.

Shapiro, James A. "Evolution without Accidents, Despite Advances in Molecular Genetics, Too Many Biologists Think that Natural Selection is Driven by Random Mutations." *Aeon*, July 6, 2023. https://aeon.co/essays/why-did-darwins-20th-century-followers-get-evolution-so-wrong.

Swinburne, Richard. *The Existence of God: Revised Edition*. Oxford: Clarendon Press, 2004.

Tanner, Kathryn. *Christ the Key*. Cambridge: Cambridge University Press, 2012.

Wainwright, Elaine. "Hear then the Parable of the Seed: Reading the Agrarian Parables of Matt 13 Ecologically." In *The One Who Reads May Run: Essays in Honour of Edgar W. Conrad*, edited by Roland Boer et al., 125–41. New York: T&T Clark, 2012.

White, Lynn, Jr. "The Historical Roots of Our Ecological Crisis." *Science* 155:3767 (1967) 1203–7. https://doi.org/10.1126/science.155.3767.1203

Wright, N.T. "Jesus and the Identity of God." *Ex Auditu* 14 (1998) 42–56.

Divine Providence and Critical Liminalities: An Ecotheopoetic Search for Post-Anthropic Subaltern Planetary Ethics

Baiju Markose[1]

■ Introduction

The lived experiences of subaltern critical liminalities demand a new perspective on divine providence. The subaltern ecological communities appropriate God's providence as an intentional practice of resistance and an unwavering undercurrent of their quotidian struggles. It is manifested in many forms of ethical expression, such as deep solidarity and distributive agency. The search for a post-anthropic subaltern planetary ethics from an ecotheopoetic vantage point draws inspiration from this conviction

1. Baiju Markose is assistant professor of theology at Trinity Lutheran Seminary in Columbus, Ohio.

How to cite: Markose, B 2024, 'Divine Providence and Critical Liminalities: An Ecotheopoetic Search for Post-Anthropic Subaltern Planetary Ethics', in EM Conradie & UL Vaai (eds.), *Making Room for the Story to Continue?*, An Earthed Faith: Telling the Story amid the "Anthropocene", vol. 4, AOSIS Books, Cape Town, pp. 111–124. https://doi.org/10.4102/aosis.2024.BK415.05

of divine providence and distributed agency affirmed by the subaltern religious wisdom. Claiming critical liminality as the new centrality, ecotheopoetics invites us to start from the middle of the happening of multiplicities. A subaltern appreciation of divine providence as a practice of resistance embodies a latent planetary ethics that reimagine human subjectivity as an emplaced subjectivity relationally integrated into the continuum of the larger ecosystem. Trusting in deep solidarity, distributed agency, and *zoe*-centered egalitarianism, subaltern planetary ethics resist any form of graded ontology in the "Anthropocene." I will start by reflecting on the poetic rendering of "Wood-Pecker" and set the platform unfolding through ornitheological reflections.

■ Wood-Pecker

Crushed in your memories,
bruised in your agonies,
I wonder how you transcend
melancholy into melodies!
In the eerie nights of solitude,
dazed in the fury wind,
drenched in the misty rain,
you might have dreamt of a winged life!
One day, you will trust your wings,
stand on your ground,
and burst out in your spirit,
to reach the limits of the sky,
and to chat with the stars.
Thank you for being with me,
and teaching us;
We journey each other home,
as we are the intertwined interbeings!
Yes, for sure,
the woodpecker is a bio-mystical relationship
between the wood and its wounds! – Wood-Pecker[2]

Poised between the earth and sky, clinging to a teak tree, evoking the image of the son of man, a woodpecker rekindled my theopoetic eco-consciousness. Driven by a determination to ignite that mystical contemplative experience, this essay takes the shape of an introspective

---

2. See Markose, *The Cross and The Peacock*, 20.

quest to uncover the potential for a theological symbolic window to be opened. The ecosystem that wraps us offers a plethora of novel symbols. These new symbols pave the way for fresh insights, providing a fresh language through which we may reimagine the divine, nature, and ourselves. The impoverishment of language is what ultimately diminishes the vibrancy of theological imaginations. Indeed, theology is an "imaginative construction."[3] Therefore, we need to fashion a new universe of symbols. The new universe of symbols will be the work of dissenting writers, poets, and artists who feel with the masses and are radically honest to the pure urge for creation. Therefore, theology must be revitalized as it explores new ecotheopoetic symbols in the "Anthropocene." This essay explores post-anthropic subaltern planetary ethics through the lens of ecotheopoetics based on the subaltern conviction on divine providence. By considering critical liminalities as the fundamental perspective, we uncover the insights and wisdom of those on the margins (subaltern) and their take on divine providence as an ecopoetic practice of resistance, which becomes the foundation for ethical imagination. The concept of critical liminality invites us to explore from the middle, rejecting the narrow confines of rigid beginnings and endings.

■ Nesting the Ecotheopoetic Imagination

I invite you to offer theopoetic attention to the poetic piece presented at the outset of this essay. The poem, penned two years ago, recounts the poignant tale of two woodpecker nestlings who tragically lost their nest during a tumultuous night of merciless storms and relentless rain. My loving life partner, children, and I endeavored to offer solace and sustenance to these vulnerable young woodpecker fledglings, only to witness their untimely demise a few days later. This heartrending event served as a profound catalyst, igniting a theological contemplation on the fleeting existence of woodpeckers, symbolic of Christ's ultimate sacrifice on the cross, unifying the concepts of wounds and wood. The haunting memory of the fledglings compelled me to re-evaluate "woodpecker" as a hermeneutical key enabling a deeper exploration of the divine woodpecker existence embodied in Jesus Christ, who hung on a tree (Acts 5:20), suspended between the realms of heaven and earth. Thus, it unveils the divine intercarnational relationship with the entirety of the world (John 3:16). It is worth noting that the historical fact of Jesus, as a skilled carpenter working with wood, further inspired me to establish this extraordinary connection between Jesus and the woodpecker, unveiling a profound and intricate bond.

---

3. Kaufman, *In Face of Mystery*, ix.

Humans are not entirely separate from nature and more-than-human neighbors. As Thich Nhat Hanh suggests, genuine neighborliness with more-than-human materiality will ignite the realization that we are inter-beings, not independent beings.[4] This awareness of interdependence emerges from a deeper understanding of divine providence. Nature is the manifesting platform of divine providence and interdependence. To appreciate divine providence, we need to find ourselves in a continuum with nature.

The sad, subtle death of the two woodpecker nestlings and the agony that brought us together made me realize that there are deep ontological roots of mutual sharing in our very existence, which we deliberately forget in the fake hustle and bustle of life's struggle. In light of this event, I tried to trace an ornitheological connection between the wood, woodpecker, wound, Christ, marginality, and liminality.

The woodpecker's existence, hanging on a tree between heaven and earth, exemplifies the "liminal" life. When we translate the symbol of the "wood-pecker" as an embodiment of liminal life and interpretive device, it is better to use a hyphen in between "wood-pecker." The hyphen in the middle is a careful record of the life of critical liminality. The experiences of in-between/betwixt problematize simultaneously the social condition of being inside and outside. Such stumbling experiences mark the lives of the marginalized. The subaltern lived experiences of Dalit people, particularly in India, punctuate a unique liminal location of "victim-insider" (*Dasyus*) and "untouchable-outsider" (*Asprsyas*) and "visibility" (*prathyaksha*) and "invisibility" (*aprathyaksha*). This is the existential dilemma of the "in-betweenness" of the subaltern, which I name as the "critical liminalities," referring to both avowal and disavowal of subaltern body in a caste-stigmatized sociality.

If marginalization results from various exclusions in society, the experience of liminality is its derivative. It is essential to distinguish between marginality and liminality. There is a marked difference between the marginality imposed by oppressive structures and the liminality one chooses as the site of resistance! Liminality is a place of openness and possibility. bell hooks says, "I make a definite distinction between that marginality imposed by oppressive structures, and that marginality one chooses as the site of resistance, as location of radical openness and possibility."[5] Critical liminality is the marginality that one chooses as the site of resistance. In that way, liminality is the creative recovery of marginality. One can voluntarily enter liminality without experiencing marginality.

---

4. Nhat Hanh, *Call Me by My True Names*, 154.

5. bell hooks, *Yearning*, 153.

However, for marginalized people, liminality is an unavoidable experience and can sometimes be suppressed by the authorities! In other words, although liminality is not necessarily an experience of marginality, every experience of marginality necessarily involves liminality because marginal space overlaps liminal space! In short, marginal experiences have two critical strands. The first is the negative connotation of being excluded by the dominant group. The second is the positive aspect of being a creative liminal space for creative engagements. I noticed that the critical liminalities of the people at the margins overlapped with the larger ecological liminalities, and the subaltern people share "ontological relationality" with the larger natural ecosystem and thereby an intentional and intensive appreciation of divine providence.

■ Theopoetic Christ: Ornitheological Entanglements

An ornitheological quest[6] for connecting Jesus Christ and the wood-pecker inspired me to consider the ecotheopoetic method as a suitable assemblage paradigm, which combines eco-consciousness, poetics, and theology, traversing the disciplinary borders intentionally. The ecotheopoetic entanglements between the liminal spacing of the woodpecker on the tree, Jesus hanging on a tree, and the lynched black/brown bodies hanged on the tree (the strange fruit)[7] metaphorically inform each other. Acts 5:30–32 accurately portrays the critical liminality, as it states, "God of our ancestors raised up Jesus, whom you had killed by hanging on a tree. God exalted him at his right hand as Leader and Savior that he might give repentance to Israel and forgiveness of sins. And we are witnesses to these things, and so is the Holy Spirit whom God has given to those who obey him" (New Revised Standard Version, Updated Edition). This is the heart of the kerygma. The betwixt experience of Christ on a tree redefines life at the margins. The notion of *homo sacer* sheds more theoretical light on this experience of the inside-outside existential dilemma. Agamben picks the figure of *homo sacer* from archaic Roman law. According to Roman law, a *homo sacer* is the individual who "has been excluded from the religious community and all political life [...] What is more, his entire existence is reduced to a bare life stripped of every right because anyone can kill him without committing homicide; he can save himself only in perpetual flight

---

6. Ornitheology is a term coined by the great theologian and birder John Stott in his book *The Birds, Our Teachers*. Ornitheological research centers around the idea that birds have something to say about truth, beauty, and a crazy-creative God. See https://www.ornitheology.com/post/ornitheology_explained [last accessed October 11, 2024].

7. See Cone, *The Cross and the Lynching Tree*, vii.

or foreign land."[8] Rejected by the law and ejected from the political community, a *homo sacer* can be killed by anyone. Because of this, *homo sacer* is exposed to the threat of death at any time. The people at the margins, especially Dalit people, live the experience of such a "state of exception." Abhijith says that the marginalization of Dalits through casteist practices has made Dalits the *Homines sacri*. As such, they are a people who are part of the society but consciously expelled by the sovereign to the margins.[9] In every case of marginality, Christ comes as an interlocutor. Through the assemblage of thought experimented with above, inspired by ornitheology, we reimagine theopoetic Christ as an active interlocutor in the ecological entanglements of the subaltern. By intertwining the metaphors of the wood-pecker, Christ, wood, and wounds, we gain a deeper understanding of the state of exception experienced by those at the margins. In this vision, Christ emerges not only as a figure of divine redemption but also as a companion in the struggle against systemic oppression and ecological degradation.

In essence, the ecotheopoetic exploration of suffering in the "Anthropocene" invites us to see God's loving care not as a passive assurance but as an active force of resistance and solidarity. It challenges us to confront injustice and ecological destruction with the same fervor and compassion embodied by Jesus on the cross. Through this lens, trust in God's providence becomes not a justification for complacency but a call for deep solidarity.

■ Ecotheopoetic Imperative of Liminality in the "Anthropocene"

Many ecotheopoetic thinkers strongly criticize reducing ecological crises to a threat to human survival. Roland Faber astutely names this fallacy the anthropic fallacy. He writes, "The anthropic principle expresses itself most prominently and obviously in the form of the anthropic fallacy in the current ecological discourse, such that to save nature from human influences is actually a strategy of human survival."[10] Rather than situating humanity anew by redefining nature as organic integrity in which humanity is relationally integrated, we often fallaciously tend to put human subjects over the nature-other. With a similar critical sense, Anna Tsing draws attention to the "anthropo-." The qualifying prefix often falls short of defining the "Anthropocene," as it fails to trace the root cause of ecological

---

8. Agamben, *Homo Sacer*, 183.

9. Abhijith, "Dalits as Homines Sacri," 41484.

10. Faber and Fackenthal, *Theopoetic Folds*, 218.

crises to exploitative socio-economic systems.[11] This reference to the unspecified humanity by using "anthropo-" blocks our attention to the patchy landscapes, multiple temporalities, and the shifting assemblages of human and nonhumans.

According to Callid Keefe-Perry, the ecotheopoetic search is qualified by the "acceptance of cognitive uncertainty regarding the divine, an unwillingness to attempt to unduly banish that uncertainty, and an emphasis on action and creative articulation regardless."[12] James Hill approaches the theopoetic search from a subaltern intersectional point of view: "A key to understanding the rich theopoetical tradition of black religious communities is how their theopoetic practices were used to inaugurate alternative worlds of anti-colonial, socio-political possibility; worlds that were often sequestered from them by the necropolitical forces surrounding them."[13] The theopoetic approach's sensibility to start in the middle, in dialogue with pluriform hermeneutic possibilities, destabilizes any forms of totalitarian interpretation. It entails an ethical and spiritual call to "always begin in the middle."[14] Deleuze formulates this new categorical imperative of eco-ethics as letting our loves be like the wasp and orchid without beginning and end, as continually in the middle, between things, interbeing, and intermezzo. Faber explicates what it means to begin in the middle.

> To begin in the middle always means to follow multiplicities in their deconstructive complexity within and without, to unsettle the boundaries and clear borders of forced identities, which are always imposed measures of the one with its power-installed abstractions of unification and division. To begin in the middle is an ethical category that activates us from the middle of the happening of multiplicities and asks us to always submerge into their middle and many folds of connectivity within and beyond, which always form under the skin of powers of unification and division and only come to life within, across and beyond the boundaries of power.[15]

To embody the liminal, we must transcend the elevated realms of unity and actively engage with the intricate dynamics within and between various elements. We must acknowledge the moments of connection between different entities and the artificial barriers that isolate them. Embracing being "in-between" requires us to adopt an intermezzo mindset, contrasting the rigid abstractions that perpetuate a sense of dominance over nature, culture, and ourselves. This mindset encourages us to embrace a more

---

11. Tsing, *The Mushroom at the End of the World*, 20.

12. Keefe-Perry, *Way to Water*, 111.

13. Hill, *A Creative Collaborative for Theopoetics*. https://artsreligionculture.org/

14. Deleuze and Guattari, *A Thousand Plateaus*, 17.

15. Faber and Fackenthal, *Theopoetic Folds*, 232.

humble and nuanced perspective. To embody the liminal is to challenge the notion of fixed identity and instead embrace a multidimensional existence akin to a river's flowing nature or the rhizome's interconnectedness.

The experience of critical liminality calls us to a new mode of knowing, being, and acting in the world. This new approach is essential for addressing subaltern histories marked by disruption, displacement, and irrecoverable loss. It is an approach that Derek Walcott, a Caribbean poet of African descent, describes as a "gathering of broken pieces." Walcott says, "Break a vessel! The love that reassembles the fragments is stronger than the love that took its symmetry for granted when it was whole."[16] For a subaltern community, assembling the shattered history together is a painful but redeeming act of subjectivity formation. Mayra Rivera names such an intellectual practice as poetics.[17] *Poiesis* means creative making, specifically an "intellectual practice attentive to events [which] shuns totalizing forms of thought and writing. It questions the search for legitimacy in genealogies and the drive to produce ontological systems, theories of the nature of being itself."[18] In such a way, poetics becomes an approach to knowledge that values creation processes from "shattered histories" and "shards of vocabularies." This approach acknowledges subversive discontinuities. This poetical longing is part and parcel of subaltern liminality, a painful process of remaking visions of corporeality from and out of pieces of shattered histories and shards of vocabulary.

■ Ecotheopoesis of Subaltern Solastalgic Communities: Divine Providence as the Grounding and Practice of Resistance

Solastalgia, as defined by Glenn Albrecht, is the experience of being homesick without leaving home,[19] caused by environmental damage. Indigenous subaltern communities like Dalits, tribal people, and Adivasis carry solastalgia in their bodies, even though they are the ones who are least contributing to ecological destruction. The marginal social location of these communities and the critical liminalities necessitate a strong trust in the itinerant divine providence amid many insecurities in their lives. God's providence is appropriated by such subaltern religious communities as an intentional practice of resistance and an unwavering undercurrent of their quotidian struggles. Deep solidarity, agential realism, and distributive

---

16. Walcott, *The Antilles*, 3.

17. Rivera, *Poetics of the Flesh*, 3.

18. Rivera, *Poetics*, 3.

19. Albrecht, "Solastalgia," 34.

agency emerge as the ethical expressions of this subaltern sensibility of divine providence. Here, I want to account for two biographical examples of the practice of divine providence from a subaltern ecotheopoetic perspective.

The Practice of Pārppu: My Grandma's Fugitive Wisdom

The memory of my maternal grandma, now fading like the gentle twilight, resurfaces with the brilliance of her distinctive words. She was one among the Dalit Christian fugitives who escaped the caste slavery in the South Travancore and migrated to the northern parts of Western Ghats in Kerala during the mid-twentieth century.[20] My grandma lived a wood-pecker (liminal) life in the casteist sociality of twentieth-century Malabar. My childhood during the last decades of the twentieth century was replete with the enchanting days spent at her side, traversing the captivating pilgrimage to her abode. The journey, adorned with the richness of fresh red soil, enveloped by the comforting presence of coconut trees, punctuated by quaint crossroads, and graced by the iconic hanging wooden bridge over the river, unfurled a vivid canvas of communion with nature. Each step etched a sacred ritual, and the nights spent on a humble mat beneath the vast night sky carved a redemptive sanctuary within my soul—a solace during moments of loneliness.

Amid the kaleidoscope of Malayalam[21] colloquial expressions generously shared by my grandmother, one term stood out with unparalleled significance—"*Pārppu.*"[22] Inflected with colloquial intonations and subaltern nuances, its importance transcended the confines of mere language. As a Dalit Christian woman navigating the intricate tapestry of Christian spirituality in twentieth-century Malabar, North Kerala, my grandma embodied an uncommon fusion of Eastern Christian Orthodoxy, Anglicanism, and the radical wisdom of subaltern Christian becoming.

Delving into the essence of *Pārppu* in my grandmother's distinctive vernacular, I discern a profound connection to Dalitity—a term encapsulating Dalit communities' marginalized existence and resistance. She used to insist we stay (*Pārppu*) with her one more day and one more day (many more), and she used this particular word, *Pārppu*, to refer to that experience of with-nessing together. I assume the term pulsates with a call for "staying"

---

20. Mohan, *Echoes of Caste Slavery in Dalit Christian Practices*.

21. Malayalam is one of the prominent South Asian languages, with a Dravidian lineage spoken in the Indian state of Kerala and the union territories of Lakshadweep and Puducherry by the Malayali people.

22. The literal meaning of the Malayalam word *Pārppu* is staying in the abode.

in authenticity, urging a radical dwelling unbound by fixity. It invites us to trust in the agential capacity of intra-actions between humans and more-than-human beings, instilling a complete faith in the divine providence and agential realism that encapsulates the fugitivity of a subaltern woman. This transformative force is both therapeutic and revolutionary.

Embedded within the framework of new materialist theory, particularly articulated by Karen Barad, *Pārppu* seamlessly aligns with the concept of agential realism.[23] Barad's insistence on distributive agency, necessitating grappling with the intricacies of existence without seeking simplistic resolutions, finds resonance in the notion of *Pārppu*. It calls us to confront the uncertainties of life with resilience and fortitude. Donna Haraway's concept of "staying with the trouble" further illuminates the profound significance of *Pārppu* in our contemporary context, demanding a steadfast commitment to addressing the challenges of our times without succumbing to despondency or indifference. Haraway writes what it means to stay with the trouble: "to make kin in lines of inventive connection as a practice of learning to live and die well with each other in a thick present [... it is the] learning to be truly present [...] as mortal critters entwined in myriad unfinished configurations of places, times, matters, meanings."[24] In the current ecological tipping point, staying with the trouble and avoiding fixed categorizations becomes paramount, echoing Barad's perspective. This harmoniously aligns with the tenets of *Pārppu*, a kind of tentacular thinking that encourages us to embrace the complexities and uncertainties of existence with an open heart and an engaged mind. In this sense, *Pārppu* is an invitation to navigate the intricacies of life without succumbing to the allure of simplistic resolutions.

Pārppu emerges not only as a call for staying with the trouble but as a manifesto for embodied resistance—a bold invitation to authentically inhabit the flux of existence, placing trust in the agential capacities of divine providence and through including more-than-human entities. It encapsulates the fugitive wisdom of Dalit spirituality, serving as itinerant hope and resilience in a world marked by marginalization and exclusion. The cinematic portrayal of Harriet Tubman in the movie *Harriet* vividly captures her unique contemplative practice of *Pārppu*, a radical act of communion with the more-than-human ecosystem trusting the divine providence for an itinerant/subversive discernment of the "present," showcasing its cross-cultural resonance as subaltern ecological wisdom.[25] This attests that another world is not merely on its way; it already exists

---

23. Barad, *Meeting the Universe Halfway*, 132.

24. Haraway, *Staying with the Trouble*, 1.

25. Vincenty, "Harriet Tubman Really Did Have Visions."

among the people at the fringes, weaving its threads through diverse cultures and practices, beckoning us to embrace its transformative potential.

Arboreal Activism: Kallen Pokkudan's Emplaced Wisdom

Kallen Pokkudan, renowned as the "mangrove man" of Kerala, dedicated his life to re-establishing mangrove forests throughout the state. His autobiography, *Kandalkaadukalkkidayil Ente Jeevitham*,[26] not only tells his life story but also sheds light on the mangrove habitats in Kerala's Kannur district and the Dalit Pulaya community to which he belongs. Given recent critical approaches to vegetality, Pokkudan's arboreal activism can be interpreted in a new context. Pokkudan, born into a Dalit Pulaya family of agricultural laborers in Kerala, spent his life planting thousands of mangrove saplings along the banks of the Pazhayangadi River in the Kannur district. By intertwining human life experiences with vegetal histories in his autobiography, Pokkudan has created a valuable collection of ecological knowledge and nature conservation principles. Recognizing that the socio-economic well-being of local communities relied heavily on the land–sea liminal ecology and its mangrove habitats, he constructed mangrove walls along the beaches of Kannur district to combat issues related to sea erosion. In this context, "Dalit eco-narrations of the self,"[27] like Pokkudan's *Kandalkaadukalkkidayil Ente Jeevitham*, expresses the autobiographical relationship between an environmental space intertwined with survival and struggle. Additionally, Pokkudan's autobiography can be seen as an experimental form of life-writing, where the subaltern subject is situated within the surrounding ecosystem of the mangrove forest and its unique ecology. "The 'self/setting' binary in life narratives can be eliminated by situating the self within the oikos, which signifies the idea of a larger home that consists of human and other-than-human actors. This emplacement establishes a continuum between the oikeion (traditionally associated with women, nature, and subalterns) and the politikon or the public sphere [...] In other words, Pokkudan's 'subaltern oiko-autobiography' reimagines the 'I' in life writing as an 'emplaced subjectivity' as opposed to an 'enclosed self'."[28]

---

26. Pokkudan, *Kandalkaadukalkkidayil Ente Jeevitham*, 1.

27. Mukul Sharma, "My World Is a Different World': Caste and Dalit Eco-LiteraryTraditions". South Asia: Journal of South Asian Studies 42:6 (2019): 1013, 1018.https://doi.org/10.1080/00856401.2019.1667057.

28. Pariyadath, *Mayilamma*, xxii, xxx.

Pokkudan highlights the historical dehumanization of Dalits within India's caste system, explaining how the Pulayas in Kerala were denied their humanity by being given names associated with animals and worms. This relegated them to a lower status in society. However, Pokkudan believes this dehumanization can be challenged and reversed, and he reflects on the potential for a shift in perspective within Indian society. He sees the Dalit vernacular in life narratives as a way to reclaim their place in society. Throughout his autobiography, mangrove is a metaphor for a relational way of being, disrupting essentialist thinking. In Pokkudan's narrative, the mangrove represents the struggles of the poor who face caste-based discrimination and poverty. Venugopal and Rangarajan rightly call Pokkudan's ecological activism "arboreal activism" and "dark ecology."[29] Pokkudan's dark ecology is grounded by its emphasis on the emplaced human subjectivity. He proposes that all reality inhabits a single plane of immanence by placing human subjectivity in the middle of the ecological entanglements. The idea of emplaced subjectivity underscores the consciousness of environmental (divine) providence and distributed agency. Pokkudan's approach is a kind of object-oriented ontology (O-O-O) and *zoe*-centered egalitarianism, which poses a strong critique against the graded ontology of casteism and other supremacies of the "Anthropocene."

Towards Subaltern Planetary Ethics

Both examples above show latent planetary ethics emerging from the heart of subaltern ecotheopoesis. The most profound awareness of divine providence experienced through the larger ecosystem, appropriated by intra-actions between emplaced human subjectivity and the more-than-human entities, serves as an epistemological basis of this planetary ethics. An object-oriented ontology that destabilizes the graded ontologies of the "Anthropocene" is another baseline of the ecotheopoetic engagement from which the subaltern planetary ethics unravels. Deep solidarity, distributed agency, and *zoe*-centered egalitarianism express this subaltern planetary ethics and politics. I propose three critical considerations as we develop a post-anthropic subaltern planetary ethics:

1. By considering the lived experience of critical liminalities and the itinerant divine providence of the marginal ecological communities as the epistemological starting point, subaltern ecotheopoetics invites us to start from the middle of the happening of multiplicities and asks us to always submerge into their middle, attending to the patchy landscapes, multiple temporalities, and the shifting assemblages of

---

29. Venugopal and Rangarajan, "Life Writing as Plant-Writing," 239–52.

humans and nonhumans. In this way, ecotheopoetics introduces a new cartography for global ecological engagements.
2. Subaltern planetary ethics evoke an ecological death (composting) of the traditional theological dispositions of the idea of a wholly transcendent God and privatized divine providence. Ecotheopoetics reimagines God as theopoetic divine, a reality that shares the same plane of immanence. It also reimagines divine providence as the grounding of planetary existence and interdependence between life forms, and it is a practice of resistance of the subaltern communities.
3. The subaltern planetary ethics should be understood as an emplacement ethics that positions human subjectivity as an emplaced subjectivity relationally integrated into the continuum of the (*oikos*) ecology and (*politicos*) politics, holding on to the ethical expressions of deep solidarity, distributed agency, and the *zoe*-centered egalitarianism against any forms of graded ontologies of the Anthropocene.

■ Conclusion

The examination of subaltern critical liminalities prompts a re-evaluation of our conception of divine providence in the face of the suffering endured by God's creatures in the "Anthropocene." For marginalized ecological communities, divine providence transcends mere theological abstraction; it becomes a deliberate act of defiance interwoven into their daily struggles. This defiance manifests through profound solidarity and shared agency, serving as the moral compass of their existence. In this context, the pursuit of post-anthropic subaltern planetary ethics emerges, drawing inspiration from the belief in divine providence and collective agency upheld by marginalized religious traditions. Through the lens of critical liminality, ecotheopoetics challenges us to perceive reality amid diversity, acknowledging the myriad perspectives and realities that shape our world. Embracing divine providence as an act of resistance encapsulates an inherent planetary ethics that reconceptualizes human identity as intricately linked with the broader ecosystem. Rooted in principles of deep solidarity, distributed agency, and *zoe*-centered egalitarianism, subaltern planetary ethics reject hierarchical ontologies' frameworks of the "Anthropocene," paving the way for a more inclusive and sustainable future.

■ Bibliography

Abhijith, T.S. "Dalits as Homines Sacri: The Politics of Inclusive Exclusion in Select Dalit Poetry." *International Journal of Current Research* 8:11 (2016) 41483–86.

Agamben, Giorgio. *Homo Sacer: Sovereign Power and Bare Life*. Translated by Daniel Heller-Roazen. Stanford: Stanford University Press, 1998.

Albrecht, Glenn. "Solastalgia: Environmental Damage has Made it Possible to be Homesick without Leaving Home." *Alternatives Journal* 32:4&5 (2006) 34+. *Gale Academic OneFile* link.gale.com/apps/doc/A161545303/AONE?u=anon~87ca3e40&sid=googleScholar&xid=d5f4f3fa.

Barad, Karen. *Meeting the Universe Halfway: Quantum Physics and the Entanglement of Matter and Meaning*. Durham: Duke University Press, 2007.

Cone, James H. *The Cross and the Lynching Tree*. New York: Orbis, 2011.

Deleuze, Gilles, and Felix Guattari. *A Thousand Plateaus*. London: Bloomsbury Revelations, 2013.

Faber, Roland, and Jeremy Fackenthal. *Theopoetic Folds: Philosophizing Multifariousness*. New York: Fordham University Press, 2013.

Haraway, Donna J. *Staying with the Trouble: Making Kin in the Chthulucene*. Durham: Duke University Press, 2016.

Hill, James. *A Creative Collaborative for Theopoetics*. https://artsreligionculture.org/.

hooks, bell. *Yearning: Race, Gender and Cultural Politics*. Boston: South End, 1990.

Kaufman, Gordon D. *In Face of Mystery*. Cambridge: Harvard University Press, 1993.

Keefe-Perry, L. Callid. *Way to Water: A Theopoetics Primer*. Eugene: Cascade Books, 2014.

Markose, Baiju. *The Cross and The Peacock*. Delhi: ISPCK, 2021.

Mohan, Sanal. "Echoes of Caste Slavery in Dalit Christian Practices." https://blog.oup.com/06July2015/echoes-of-caste-slavery-in-dalit-christian-practices/.

Nath Hanh, Thich. *Call Me by My True Names*. Berkeley: Parallax Press, 1999.

Pariyadath, J. *Mayilamma: The Life of a Tribal Eco-Warrior*. Translated by S. Rangarajan and S. Varma. Hyderabad: Orient Black Swan, 2018.

Pokkudan, Kallen. *Kandalkaadukalkkidayil Ente Jeevitham*. Compiled by S. Paithalan. Kannur: Media Magic, 2002.

Rivera, Mayra. *Poetics of the Flesh*. Durham: Duke University Press, 2015.

Samantha, Vincenty. "Harriet Tubman Really Did Have Visions." *Oprah Magazine* (2 November 2019). www.oprahmag.com/entertainment/tv-movies/a29664794/harriet-tubman-visions-movie/.

Sharma, Mukul, "My World Is a Different World': Caste and Dalit Eco-Literary Traditions". South Asia: Journal of South Asian Studies 42:6 (2019): 1013, 1018. https://doi.org/10.1080/00856401.2019.1667057.

Tsing, Anna Lowenhaupt. *The Mushroom at the End of the World: On the Possibility of Life in Capitalist Ruins*. Princeton: Princeton University Press, 2015.

Van Wieren, Gretel. *Restored to Earth: Christianity, Environmental Ethics, and Ecological Restoration*. Washington, DC: Georgetown University Press, 2013.

Venugopal, Varna, and Swarnalatha Rangarajan. "Life Writing as Plant-Writing: Arboreal Encounters in Kallen Pokkudan's *Kandalkaadukalkkidayil Ente Jeevitham* (2002)." *Lagoonscapes. The Venice Journal of Environmental Humanities* 2:2 (2022) 239–52. https://doi.org/10.30687/LGSP/2785-2709/2022/04/004

Verschuuren, Bas, et al. *Sacred Natural Sites: Conserving Nature and Culture*. London: Routledge, 2010.

Walcott, Derek. *The Antilles: Fragments of Epic Memory*. New York: Straus and Giroux, 1993.

The Challenge of "Angry Weather": Planetary Perspectives on the Providence of God from Life Lived Down Under

Clive Pearson[1]

■ Situating the Organizing Question

The theological question shaping this anthology is played out in multiple contexts, expressed in the following form: "How could the suffering of God's creatures in the 'Anthropocene' be reconciled with God's loving care?" Thus it situates itself at the intersection of the doctrine of providence and the theodicy problem. It does so at a point in time where the former doctrine has been subject to extensive critique in the public domain.

---

1. Clive Pearson is a senior research fellow in the Centre for Religion, Ethics, and Society, Charles Sturt University, Australia. He is also a research fellow in the Department of Religion and Theology at the University of the Western Cape.

How to cite: Pearson, C 2024, 'The Challenge of "Angry Weather": Planetary Perspectives on the Providence of God from Life Lived Down Under', in EM Conradie & UL Vaai (eds.), *Making Room for the Story to Continue?*, An Earthed Faith: Telling the Story amid the "Anthropocene", vol. 4, AOSIS Books, Cape Town, pp. 125–148. https://doi.org/10.4102/aosis.2024.BK415.06

Among the most damning is Susan Neiman's assertion that "[p]rovidence is a tool invented by the rich to lull those whom they oppress into silent endurance."[2] Theories of theodicy are likewise subject to complaint on the grounds that they appear to prefer abstraction to actual instances of suffering and menace.[3] In the liminal conditions of the "Anthropocene" they are vulnerable to the ecotheoethic critique that lies behind the contribution by Baiju Markose: theodicy tends to be limited to the suffering and survival of that which is all too human.[4]

The reference to the "Anthropocene" suggests a very different kind of context. At one level, it situates us as a species (irrespective of where we live)—among other species—within a setting of the planet's deep history conceived in geological terms. It signifies a rupture to the Earth System which is being most obviously experienced in and through climate change.[5] Some considerable care is necessary here. Through his engagement with leading Earth System scientists, the Australian philosopher Clive Hamilton has argued that the discourse around the "Anthropocene" is not immune to a series of category mistakes. This rupture should not be confused with "the human environmental impact on the landscape [or] the Earth's surface." It is a radical breach that effects the Earth System "as a totality" in all its intersecting complexity.[6] The "Anthropocene" represents the possibility of a new epoch in the life of the earth—or, at the very least, "an invaluable descriptor in human-environment interactions."[7]

In a rather blunt way, Hamilton warns against the common tendency of seeking to interpret this rupture though the lenses of "the old disciplines" that belong to the previous epoch—the Holocene. In these circumstances, theology (and its doctrinal claims) finds itself in the most uncomfortable of positions: the "Anthropocene" poses the awkward question of whether this discipline can readily migrate from one epoch to another.[8] Elizabeth Pyne rightly asks how we can we tell the story in ways that are in continuity with the core claims of the Christian tradition.[9] Can this talk of rupture be readily reconciled with Nicola Hoggard Creegan's invoking of evolution[10] or the

---

2. Neiman, *Evil in Modern Thought*, 182.

3. See Pearson, "Theodicy."

4. See Markose's essay in this volume.

5. See Hamilton, *Defiant Earth*, 1-35.

6. Hamilton, *Defiant Earth*, 18-20.

7. Witze, "It's Final!"

8. See Pearson, "Is it Time?"

9. See Pyne's essay in this volume.

10. See Creegan's essay in this volume.

importance Ernst Conradie attaches to the lingering legacy of the (increasingly ambiguous) Noah covenant?[11] The Holocene was one in which there were climatic variations but which still allowed cultures, civilizations, and faiths, and now extinct and endangered species, to flourish. Is that still the case?

The planet-wide nature of the "Anthropocene" and its impact on the whole of the Earth System cannot be downplayed. In an age that is suspicious of master-narratives that seek to "order and explain human experience," the "Anthropocene" proposes "an emerging narrative of humanity as a whole."[12] That is another claim that needs to be handled with great care: there are consequences to this way of thinking that will demand scrutiny, especially in the area of causality. Whether origins and responsibility will ultimately be more important than the present and future predicament that all of creaturely existence will need to negotiate is a doubtful point. This emphasis on the whole, all of us, is not a claim that does away with the significance of local contexts, however—contexts in which talk of providence and theodicy play themselves out in real-life predicaments.

In this instance, the planetary perspective is interpreted from the particular context of Australia. There is much ambiguity in this option. On what grounds is it legitimate to heed a voice on the subject of providence emanating from a land once regarded as "the most secular under heaven"?[13] The case against is further compromised through Australia's record: it is one of the world's leading carbon emitters per capita as well as being "the biggest exporter of coal and liquefied natural gas."[14] Rather than allow room for deference to a theology of providence, it has taken an inverse pride in being designated a "lucky country" blessed with natural resources to be exploited.[15]

There is another side to this coin, though, which warrants inquiry into the merits (or otherwise) of providence. The prospects for the suffering of God's creatures in this corner of the world is high. The noted climatologist Michael Mann has described Australia as "the worst continent to live on for climate change."[16] It is the planet's driest continent; the south-east of the country is one of the three regions most subject to fire.[17] The longitudinal

---

11. See Conradie's essay in this volume.

12. Hamilton, *Defiant Earth*, 76.

13. Breward, *Australia*.

14. Paul, *Australia*, 20.

15. Horne, *The Lucky Country*.

16. Paul, *Australia*, 20.

17. See Gergis, *Sunburnt Country*, 192–96.

and latitudinal of location and topography of Australia make it vulnerable to extremes in heat, drought, and flood—so much so that Joëlle Gergis contends that such events are "a characteristic of our climate: they are a defining feature of our national identity."[18] The impact of a changing climate in the "Anthropocene" has been equated with the "natural variability Australians have always had to live with now being 'on steroids'."[19]

Gergis argues that Australia finds itself being in "dangerous territory,"[20] an "age of consequences."[21] Will its levels of endemism and biological megadiversity—some of "God's [likely] suffering creatures"—be conserved?[22] The east coast fires of 2019 set an alarmingly stark record of the loss of creaturely life—other species—in a single, complex event.[23] Will the conundrum of how to manage the reduction in carbon emissions and the economy rise above the status of being a "political hot potato"?[24] Writing in the field of a political economy in the "Anthropocene," Erik Paul has made a plea for Australia to commit itself to "planetary realism."[25] The questions that surface on account of this ambiguity and uncertainty with regards to the future prompt this foray into a reading of providence.

■ Implicating Providence

In 2020, the climate scientist Friederike Otto published her monograph on *Angry Weather*. The subtitle reads *Heat Waves, Floods, Storms and the New Science of Climate Change*.[26] In the wake of a series of extreme events—including the east coast fires in Australia the previous year—the title and its personification of the weather were designed to attract. In some ways, it captured the way in which Hamilton describes how a "defiant earth"[27] is "fighting back."[28] Otto and Hamilton were seeking to capture the mood of a planet that is now more likely to be injurious to the life that the human species (wherever any of us might live) has come to assume.

---

18. Gergis, *Sunburnt Country*, 185.

19. Gergis, *Sunburnt Country*, 220.

20. Paul, *Australia*, 20.

21. Gergis, *Sunburnt Country*, 213-70.

22. See Gergis, *Sunburnt Country*, 236-44; Hughes, "Climate Change and Our Life Support System."

23. Ward et al., "Impact."

24. Gergis, *Sunburnt Country*, 250-54.

25. Paul, *Australia*, 1-8.

26. Otto, *Angry Weather*.

27. Hamilton, *Defiant Earth*, 47-48.

28. Hamilton, "Forget Saving the Earth."

Hidden away in Otto's subtitle, though, was a reference to the "new science" of climate change. Here she was not referring to the now standard kind of climate science that lies behind a succession of assessment reports emanating from the Intergovernmental Panel on Climate Change (IPCC). The emphasis was on the word "new": Otto was drawing attention to an evolving discipline called event attribution science. This term had first been mentioned in 2011 in a report on "The State of Climate" published by the American Meteorological Society, although there had been prior intimations of such.[29] It has come to mean the attempt to determine the extent of the role a warming atmosphere might play in the unfolding of extreme weather events and which might not be able to explained through recourse to natural variations.

It is not, as such, a branch of knowledge that might immediately attract theological attention. The familiar tropes of a doctrine of providence—*conservatio*, *concursus*, and *gubernatio*—do not feature in the statistical graphs and collecting of meteorological data that seek to explain the occurrence and impact of extreme events. In this respect, the new science is at a substantial remove from the interpretations that Peter Thuesen discerned in the bids to understand the intersection of divine government and the natural laws of science that accompanied the "mighty rushing winds" that have wreaked such havoc along tornado alley in the United States.[30]

In due course, these awe-inspiring storms would become subject to a technology of precise description, as well as prediction as to their likely pathway. They retain an element of mystery and, in certain quarters, can still attract sermons, prayers, media comment, and confessional asides as to how God is to be conceived in such acts. Writing in his *Tornado God*, Thuesen found it necessary to examine how the doctrine of providence had manifested itself in response to these localized cataclysmic events: this theme has been associated with these "furious" winds from the settlers' first experience of them (as they moved westwards) through to the contemporary challenge of the "Anthropocene" and climate change.

The historical invocation Thuesen discerned of this doctrine is at odds with its seeming neglect from the twentieth century onwards. Writing on *Purpose and Providence*, Vernon White reflected on a loss of confidence in this domain "that stalks our literature, philosophy, and our religion." It is like

---

29. See Peterson, "Extreme Events"; Hu, "Attribution Science."

30. See Thuesen, *Tornado God*.

a "ghost-writer" in any attempt at seeking to "make sense of life."[31] The feeling of being "haunted by providence, even when it seems impossible to believe" is itself a reflection of a disenchanted world, the loss of transcendence, and the seeming untenability of master-narratives which the doctrine seemed to assume. In a variation on a theme, Julian Gutiérrez noted how the "beginning of the twentieth century signalled a decline in the doctrine of providence."[32] That relative absence is reflected in this anthology by the frank admissions made by Gloriose Umuziranenge and Eraste Rukera in Rwanda and Elizabeth Pyne, among others in this volume, that overt talk of providence is not to be found or "has not been atop my list of theological priorities." And yet it remains—sometimes hidden from view, sometimes "implicit and inchoate," but nevertheless necessary to the schema of Christian life and practice.

That Thuesen should have discovered so much thinking on providence in the wake of a weather event is not too surprising. David Fergusson has described how this theological *locus* is a rather "capacious" doctrine. Indeed, "[i]ts scope includes the order of nature, the direction of history, the way in which the lives of persons are subject to divine guidance, the problems of evil and suffering, the language of politics, the constructions that we place upon our individual life stories, and the final outcomes of nature and history."[33]

It is a broad brief that "has been widely deployed." In terms of weather and, by extension, climate change, the line of connection arises out of the very nature of life itself. Sigurd Bergmann rightly notes that "[w]eather belongs to the essential conditions of our bodily life." The weather "impacts us all, in every place and at every moment."[34] It "simply surrounds us and embraces us." The weather is "fundamental for every living being which breathes in air."[35] In its role of being Fergusson's "sequel to creation,"[36] the doctrine of providence is played out within the conditions of life made possible (or not) by the weather.

Through the lens of providence, Thuesen found himself having to report on popular opinion as well as sustained theological enquiry into these violent storms. Was the "howling tempest" that ushered in such "dreadful

---

31. White, *Purpose and Providence*, 1.

32. Gutiérrez, "King of Glory?" 57.

33. Fergusson, *Providence*, 1.

34. Bergmann, *Weather*, 1.

35. Bergman, *Weather*, 4.

36. Fergusson, *Providence*, 1.

calamity" a judgment of God?[37] Was the "deliverance" of some who survived "the banalities" of this "mysterious power" a sign of a beneficent providence?[38] Or should this agency of destruction be seen as an intimation of the "limitation" of a seemingly "do-nothing" God?[39] With the advent of a burgeoning meteorological science, the outstanding point of tension that surfaced would be the distinction between a general and a particular providence. It might still be able to appropriate the themes of *conservatio*, *concursus*, and *gubernatio* in a more generalized, almost ahistorical way, while abstaining from the particularity of a divine "finger" in specific instances. In an age of increasing uncertainty, Thuesen wondered if "agnosticism and stoicism" are the only plausible responses.[40]

None of these kinds of considerations are to be found in Otto's work on angry weather. The language of providence is absent. What is perhaps present are the intimations of an unacknowledged secular equivalent. That this possibility might be hinted at should come as no great surprise. The impact of events that signify climate change has led to calls for a secular theodicy[41] and eschatology.[42] The deflationary case that Christine Helmer makes for "not thinking about the doctrine of providence anymore" concludes with an account of a creation in distress. She argues that the crisis of modernity has arisen out of the way in which theology justified a linear progression of history under the influence of a reading of providence which has become dystopic. Helmer claims: "The doctrine once used to uphold God's sovereignty has become misused to legitimate the progress of history with its nationalism, white supremacy, sexism, and economic colonialism."[43]

The line between providence and theodicy is becoming blurred "as modern reality spirals out of control." Helmer's response is to turn towards Luther's understanding of the hidden God and its resistance to being subject to the very categories upon which a doctrine of providence seemingly depends. Its place must be taken by the intimations of a true humanity embodied in the response to a set of questions based on Matthew 25:34–35. Among those she puts the far-from-comfortable challenge:

37. Thuesen, *Tornado God*, 71.

38. Thuesen, *Tornado God*, 74.

39. Thuesen, *Tornado God*, 117, 128.

40. Thuesen, *Tornado God*, 87.

41. Hamilton, "Theodicy," 233–38.

42. Rothe, "Governing the End Times?" 143–64.

43. Helmer, "Providence," 93.

"Where were you [...] when I became extinct?"[44] Writing on behalf of his low-lying islands, the Tuvaluan climate activist Maina Talia scans the global horizon and asks, "Am I not your *tuakoi*—your neighbour?"[45]

In the various scenarios that Otto examines, there is no explicit hint of this parabolic cry. There is, nevertheless, a profound interest in the implications of extreme weather events and the role of anthropogenic agency. Otto examines issues of responsibility, questions of justice, and the complicity of corporations.[46] It is not an overt expression of a conventional rendering of providence: in a rather immanental, secular way, there is an obvious concern for what lies ahead in the not-too-distant future as well as a desire to care for those most affected and to conserve, mitigate, and adapt as necessary. There are here, albeit in a different guise, two of the substantive notions Fergusson identifies with the noun providence— foresight and provision.[47]

■ Weather Attribution Science

Otto is one of the lead climatologists in an international network called World Weather Attribution that was founded in 2015. The publication of her monograph had been foreshadowed independently by Chris Funk in his work on how climate change contributes to catastrophes.[48] Through a mix of climate data, economic assessment reports, and personal narrative, Funk set about the task of demonstrating how climate change is affecting particular societies now and why that is happening. The burden of his argument was to overcome the tendency to think of climate change according to a model of a bathtub warming gradually everywhere at the same rate. Funk's alternative model was of a seesaw that is likely to swing up and down violently.[49] It comes as a shock to particular places at specific times.[50]

The timing of such work coincided with the more frequent usage of the descriptive label of "climate emergency," first coined in the course of a rally in Melbourne in 2009.[51] It has become increasingly common to cite climate

---

44. Helmer, "Providence," 93–95.

45. Talia, "*Tu'akoi?*"

46. Otto, *Angry Weather*, 109–90.

47. See Fergusson, *Providence*, 298.

48. See Funk, *Drought, Flood, Fire*.

49. Tasoff, "Drought, Flood, Fire."

50. Funk, *Drought, Flood, Fire*, 11.

51. See Jilly, "Climate Emergency."

change alongside instances of excessive heat, rainfall, and prolonged drought. The natural disasters that unfold are generating other fresh terms like "disaster brain" that accompany the psychological fallout, especially among children encountering firestorms and repeated floods.[52]

Both Otto and Funk are exponents of this new development in climate science. Its foundations, Otto argues, have been laid for a peer-reviewed methodology of determining "whether climate change has fundamentally changed the rules of specific weather events, played only a minor role, or been wrongfully blamed."[53] Otto has recognized that not all weather events can be blamed on climate change:[54] she was wary of news media headlines that might too quickly convey such an impression.

In the pursuit of this task, Otto and Funk were stepping outside the convention that climate scientists had observed of not wanting to comment on specific events. The more usual practice has been to reflect more globally and across a greater extent of time. The intersection between climate and real-time weather was thus not blurred. Otto was not unaware of criticisms that might then be laid against the aims of event attribution science;[55] the critique is made especially on the grounds of its reliance on a probabilistic or risk-based method.[56] Many of the studies carried out in the name of a climate-change attribution science have been submitted for publication in peer-review journals after the event. Otto noted that the report of the IPCC for 2013 allowed for attributing "specific weather events to change."[57]

Its method was inclined to "proceed in four steps: (1) measuring the magnitude and frequency of a given event based on observed data, (2) running computer models to compare with and verify observation data, (3) running the same models on a baseline 'Earth' with no climate change, and (4) using statistics to analyze the differences between the second and third steps, thereby measuring the direct effect of climate change on the studied event." Otto contends that in this way, attribution science is seeking to address the question, "did climate change play a role?" in specific extreme events "within the news time frame—so within two weeks of the event."[58]

---

52. Newby, "At Age Eight."

53. Otto, *Angry Weather*, 18–19, 59–75, 189.

54. Otto, *Angry Weather*, 125.

55. Otto, *Angry Weather*, 70, 73.

56. Olsson et al., "Ethics."

57. Otto, *Angry Weather*, 52.

58. Sneed, "Extreme Weather."

The Challenge of "Angry Weather"

In the instances under examination, causality is not so much a question of who or what economic system was responsible: these things are not ignored, but it is not the particular focus of this kind of research. Causality here had to do with the atmospheric, meteorological conditions that were at work that would lead to such intense and dangerous events. In terms of the weather itself, the perceived need was to consider the nature of the connection between cause and effect. In order to evaluate such, Otto argued that it was first essential to find out what had happened and reconstruct the event.[59]

For his part, Funk had initially set out to be what he called a "drought detective," which led then into his subsequent vocation of seeking to determine the possible role of climate change in extreme weather events. Those instances were no longer the "fingerprints" of such change, but rather more like a "slap [...] exacting a terrible toll." Funk's intention was to provide a "climate change toolkit."[60] His work effectively took a step back from particular examples as he too sought to consider lines and levels of causality for intense events. Although the purpose was always to sound a note of urgent warning on how the delicate balance of this fragile planet was being undermined, Funk's study was essentially retrospective. The fires, floods, droughts, and heat waves in East Africa, North America, and Australia had already long run their course.

The narrative Otto laid out was rather different in tone and style. The reader was caught up in the midst of events as they were brewing and then playing themselves out. The team she was leading was striving to predict the role of climate change in the intensity of event as it was happening.

■ Endgames

In the background to these works lay the research of a team of Earth scientists led by Canberra-based Will Steffen on what they described as the trajectory towards hothouse Earth.[61] Steffen and his associates acknowledged that there had always been variations in climate. The metaphor employed to describe their occurrence was of a braided river that meandered from one side of a valley to the other. As they considered the trajectories of the Earth System, Steffen's team concluded that we—that is, the human race as a whole, irrespective of questions of causality and accountability—had moved into an altogether different valley: "we"

---

59. See Otto, *Angry Weather*, 32–34.

60. Funk, *Drought, Flood, Fire*, 5, 7.

61. Steffen et al., "Trajectories."

cannot return to the valley that we have left.[62] These Earth System scientists provided graphs and statistics to support their hypothesis. The possibility of creating a stabilized earth pathway would only be achievable if humanity acted with a "coordinated, deliberate effort to manage our relationships with the rest of the Earth System."[63] That pathway will not be the same as what has transpired in the Holocene. It cannot be brought into being without appropriate human stewardship.[64] In the wake of the 2021 United Nations Framework Convention on Climate Change Conference of the Parties (COP26) in Glasgow, Bill McGuire, an expert in geophysical and climate hazards, concurs. The whole human race no longer finds itself dwelling on a planet in the so-called Goldilocks zone, which was a time and space that allowed cultures and faiths to flourish.[65]

There is a significant implication for reflection on providence. It is evident that we no longer inhabit, in some fairly basic respects, the conditions of life in which the theme of weather in the Bible was invoked. Bergmann had identified how the scriptural text had suggested that weather might be seen as a sign of God's continuing providence to humankind in general ("the rain falls on the just and the unjust"—a "gift") or serve as "a barometer of God's love and morality."[66] Otto's theme of angry weather might at face value seem to intimate how the weather might become "a symbol and physical expression of God's power in and over creation."[67] That power could be deployed for the sake of "vengeance and punishment" or "as a revelation of God's love and care."[68]

Evidence for how this kind of understanding might have played out even in recent times can be seen in Thuesen's *Tornado God*. Over the course of several centuries, theologians contemplated the awesome, unpredictable, seemingly numinous quality, the *mysterium tremendum*,[69] of the destructive powers of tornados that invoked biblical texts like those to do with the "fierceness" of "the whirlwind of the Lord"[70] and "the wrath of God."[71]

---

62. Steffen, "The Anthropocene."

63. Steffen et al., "Trajectories," 8259.

64. Thomas, *The Anthropocene*, 170–71.

65. McGuire, *Hothouse Earth*, 17–20; Funk, *Drought, Flood, Fire*, 16, 21–39.

66. Bergmann, *Weather*, 5–6.

67. Bergmann, *Weather*, 82.

68. Bergmann, *Weather*, 82.

69. Thuesen, *Tornado God*, 8.

70. Thuesen, *Tornado God*, 11–34.

71. Thuesen, *Tornado God*, 8, 17, 24.

For those who might be spared, the power of these mighty storms led invariably to questions of theodicy and a consideration of the providence of God. Sermons in the aftermath of such storms struggled to make sense of what had happened and where God could be found in the cataclysm. On what grounds might a member of a congregation nevertheless proclaim "my God is a merciful God, a compassionate God"?[72]

To make that kind of correspondence is to ignore the extent to which weather has acquired what Bergmann has called "a technological writeability."[73] Through the emergence of meteorology, the weather has become subject to instruments. Through the forecasting industry, it has become a commodity. Bergmann concludes that weather has lost the "spiritual aura" it once possessed.[74] McGuire's usage of the language of hothouse Earth is far removed from that past. His terminology, like that of Steffen, is not a rhetorical flourish. It is used "formally, in a definitive sense, to describe the state of our planet in the geological past when global temperatures have been so high that the poles have been ice-free."[75] The distinction is made between a hothouse state and hothouse conditions, which McGuire observes "are already becoming far more commonplace, and fast becoming the trademark of our broken climate."[76]

Neither Steffen nor McGuire can provide us with much in the way of words of comfort. In June 2020, Steffen was one of a number of top climate scientists who warned that "nine of the fifteen global climate tipping points that regulate the state of the planet have been activated." The window for "intervention time" had narrowed: Steffen advised that "civilizational collapse" was now more likely.[77] Writing in the wake of the hottest year on record, 2023, a team led by William Ripple examined the "vital signs" and warned the "time is up": "[l]ife on planet Earth is under siege" and a "worldwide societal breakdown is feasible."[78]

■ Intersectionality

That earlier reference to a homogenizing "we" is itself liable to critique. It is so especially when it is used to invoke the designation of this new era as

---

72. Thuesen, *Tornado God*, 30.

73. Bergmann, *Weather*, 73, 128.

74. Bergmann, *Weather*, 121.

75. McGuire, *Hothouse Earth*, xv.

76. McGuire, *Hothouse Earth*, xv.

77. See the report by Moses, "Collapse of Civilisation."

78. Ripple et al., "Entering Uncharted Territory," 841, 850.

the "Anthropocene"—the accidental term first deployed by Paul Crutzen[79]—but much more recently used more formally by the International Commission on Stratigraphy of the International Union of Geological Sciences (IUGS) led by Jan Zalasiewicz.[80] The work of the commission was designed to determine whether we have left the Holocene period (during which the Christian faith and human culture flourished); does the evidence for a geological golden spike testify to a new epoch—the "Anthropocene"—which bears the marks of human activity? Has humankind become a telluric force, a geological actor? The commission proposed that this transition happened in 1952 on a variety of grounds but principally through the presence of plutonium residue in the Earth's rocks.[81]

Some care needs to be exercised with the use of this terminology and its tendency towards universalizing humanity. In no small measure, the problem stems from the potential for the naming of a geologic period—and then applying it to the Earth System—in a way which implicates the whole of humankind. The key word becomes *anthropos*. Julia Thomas, Mark Williams, and Zalasiewicz use the term to reflect how "human beings are both creatures with a deep past inscribed in rock layers and ice cores, and also a new and overwhelming force on the planet." They are so as a "collective unit [...] and a multitude of disparate individuals and societies."[82] The tension they see as needing to be negotiated is one of the relationship between local and universal cultures. It expresses itself in the recognition that the "Anthropocene" "names, on the one hand, the planetary predicament faced by all beings and, on the other, the uneven responsibilities and experiences that divide us."[83]

The case against homogenizing the "us" is made very vividly by A.M. Ranawana in her liberation theology of rage that seeks to address "the bleeding wounds" of poverty and ecological damage.[84] The "Anthropocene" does not feature in the case she makes for justice on behalf of communities

---

79. Benner et al., *Paul Crutzen*.

80. Thomas et al., *The Anthropocene*, vii–viii.

81. On March 20, 2024, the parent body, the IUGS rejected the proposal made by the Anthropocene Working Group. It did so through a contested voting process where some members who had exceeded their term of membership voted in the negative. As chair of the working group and the relevant subcommission, Zalasiewicz resigned and declared his intention of codifying the "Anthropocene" through "another means." See Witze, "It's Final!" Zalasiewicz and others had previously addressed the "fundamental questions" / "the still-relevant challenges" / "the Anthropocene." Writing in *The Conversation* under the lead name of Simon Turner, he identified what critics within the subcommission had overlooked and declared "why it really should be a new geological epoch."

82. Thomas et al., *The Anthropocene*, 112.

83. Thomas et al., *The Anthropocene*, 115.

84. Ranawana, *Liberation for the Earth*.

in the Global South and Indigenous voices over and against the extractive and exploitative Global North, but the sentiment is clear. Can a theology of climate change that seeks to engage with the "Anthropocene" and planetary change ignore the inequalities in carbon emissions and the intersectional legacies of gender, race, and generations? To her list could be added other species, indeed the web of creation on which humankind itself depends. Through a focus on Indigenous and subaltern voices, Lily Mendoza and George Zachariah have overseen a project designed to decolonize ecotheology.[85] In his role as human geographer, Mike Hulme advises in no uncertain terms that "climate change is not everything" and should not be allowed to obscure multiple injustices.[86] Kathryn Yusoff is even more direct: she binds the emergence of the "Anthropocene" to what she describes as a "White Geology" inasmuch as the discipline does not reflect "histories of colonial earth-writing" and the "erasures" of Black experience.[87] In this volume, Markose invokes a "subaltern planetary ethics."

For all their merit, the difficulty with such texts is their failure to engage with the rupture to the Earth System and its impact on climate. That lack of intention serves as an invitation to consider what might then be a broader perspective. Writing on the "Anthropocene," Hamilton, Christophe Bonneuil, and François Gemenne employed a rather helpful typology.[88] They have discerned three broad discursive fields of knowledge in which the word "Anthropocene" is operating: the first, the original, is geology; the second is the emerging discipline of Earth System science and through it into climate science—that can be seen in the work of Tim Lenton,[89] one of Steffen's co-authors; the third is the social sciences, which in this matter can include ethics, history, religion, literature, art, and music as well as assorted texts to do with living and dying in the "Anthropocene." The rhetoric of the "Anthropocene" is now widely distributed: it exists even if, and when, one discursive field elects not to use it. In terms of the geopolitics of climate change—and the future of both the human and nonhuman species— it is the second field, the Earth System sciences, that is critical.

The critique is usually waged along the lines of causality and responsibility for redress. Not surprisingly, it attracts the rhetoric of climate justice, the impact of colonialism, patriarchy, speciesism, and capitalism. The intersectionality between climate change and the many other issues that

---

85. Mendoza and Zachariah, *Decolonizing Ecotheology*.

86. Hulme, *Climate Change Is Not Everything*.

87. Yusoff, *Billion Black Anthropocenes*, 2, 23-63.

88. Hamilton, *The Anthropocene and the Global Environmental Crisis*, 2-4.

89. Lenton, *Earth System Science*, 74-90.

attract the call of justice—what the Oxford philosopher Henry Shue describes as "compound injustice"— is another broad field.[90] The stakes are high.

Hamilton is adamant: the three types should not be confused. Geology and the Earth System sciences are not the same as the social sciences. It is true that some ecomodernists like Erle Ellis do speak of a "good Anthropocene" and do so in a way that is rather worrying: they assume technological developments that will benefit "most" of us.[91] They do not address the question of who is excluded from that "us"—but that is a position that is not widely commended. The political economist of climate change Simon Dalby poses the question whether the "Anthropocene" will be good, bad, or ugly and concludes it will be ugly.[92] It is simply not accurate to imagine that those who delve most deeply into the rupture that the "Anthropocene" has wrought on the Earth System disregard historical and present injustice and threat. The most cursory of readings of the highly regarded journal, *The Anthropocene Review*, readily reflects such.[93] Could it be that a theology of providence furnishes a way of overcoming this antagonism? Its three-fold tropes hold together (a continuing) creation, the unfolding of history, and teleological purpose.

■ History Types

In their multidisciplinary work on the "Anthropocene," Thomas, Williams, and Zalasiewicz argue that humankind now finds itself in a "predicament" (rather than a problem).[94] They do not seek to place one type of discourse over against the others; they argue that what is required is "networks of knowledge, all focused on the reality of the Anthropocene but using their own lenses."[95] In a similar vein, Amitav Ghosh observes that now is not the time for siloed knowledge.[96] In a rather helpful way, Thomas, Williams, and Zalasiewicz identify how these often "difficult conversations" arise out of our differences in scale and alternate interpretations of causality.[97]

---

90. Shue, *Climate Justice*, 37-42.

91. Hamilton, *Defiant Earth*, 21-27, 66-72; "Theodicy"; Asafu-Adjaye, "An Ecomodernist Manifesto"; Ellis, *Anthropocene*.

92. Dalby, "Framing the Anthropocene."

93. Pearson, "Climate-Driven Losses."

94. Thomas, *The Anthropocene*, 3.

95. Thomas, *The Anthropocene*, 5.

96. Ghosh, *The Great Derangement*, 3.

97. Thomas, *The Anthropocene*, 13-16.

Earth System scientists deal with "hyper time," longevity, magnitude, and a "data collection of colossal proportions." Social scientists are more likely to be responding to how "human communities suffer or celebrate the Earth's changes with varied ecologies and cultural systems" in which they are "nested."[98] The distinction that is then made between history in the "Anthropocene" and environmental histories provides a very necessary sense of perspective for situating a doctrine of providence. Its markers are four-fold: the first is its "unprecedented newness—[w]e have always lived in our environment: never before have we lived in the Anthropocene." The second is its scale, where its form is global, "with an eye to deep time and to the extraordinary, recent acceleration in human prodigality." The third is especially significant: history in the "Anthropocene" is responding to a "different science: Earth System science rather than ecology." And finally, there is the need to "contend with a radically new concept of human agency as a systemic planetary force."[99]

Something of this transition in the understanding of history is evident in the writings of Dipesh Chakrabarty. At one stage in his career, he was a postcolonial historian and critic. His line of approach altered while living in Canberra during the devastating bushfires of 2003. Now he found himself beginning to think for the first time about "the human as a geological agent, whose history could not be recounted from within purely humano-centric views." It led him to becoming familiar with what he describes as "the polysemic category" of the "Anthropocene." Irrespective of whether or not the geologists formalize the name, Chakrabarty was conscious of how the planetary system has been radically altered.[100]

Otto's angry weather is a symptom of such urgency and rupture. The emphasis she places on specificity amid the global is likewise reflected in the work of Ashley Dawson on extreme cities. These cities are extreme not in terms of size but rather in their capacity "to weather the storms that are bearing down upon humanity." Dawson makes a break from the common practice for this threat to humankind to be assessed on a global scale and in the future tense. That practice is at a remove from the specificities of where most people live or are likely to live. Dawson identifies the dilemma that Otto has been seeking to overcome: "Weather events, the day-to-day expression of climate, do not occur at a global or hemispheric scale. Severe storms, flood, heat waves, droughts and perhaps most important,

---

98. Thomas, *The Anthropocene*, 5–7.

99. Thomas, *The Anthropocene*, 125–27.

100. Chakrabarty, *The Climate of History*.

daily fluctuations in temperature only occur in an experiential way at the scale of religions."[101]

Dawson's work should be seen in the light of the emerging corpus of writings on the city and the coming crisis of climate change and the "Anthropocene."[102] It is now recognized that most people in the "Anthropocene" who will be affected by the impact of climate change will live in cities; they also happen to be the greatest producers of carbon emissions. In terms of a notion of providence, what becomes intriguing is the way in which cities concentrate the activity and creativity of the human species[103]—and hence the potential for running contrary to (or in compatibility) with the divine *concursus*.

■ Revisiting Providence

The burden of this rendering of the "Anthropocene" is that humanity now finds itself inside an Earth System that is no longer the one in which its cultures and religions flourished. The scale of that change is forcing a fresh understanding of what constitutes history and the need to embrace deep time. It is far from clear (if ever it was) as to "what is God up to" or "where is God to be found." The rise of an event attribution science and conceptions of angry weather intensifies the reception experience of those left wrestling with the legacy of biblical narratives to do with storms, whirlwinds, and floods, to which could be added fires and the menace of the silent killer, extreme heat.

The standard practice in discussing providence is not to begin with that which is particular. It is a recurring theme.[104] The preference is to opt for a more general understanding of providence along the lines of *conservatio*, *gubernatio*, and *concursus*. The "hinterlands" of the doctrine in the philosophy of Plato and the Stoics (as demonstrated by Gijsbert van den Brink)[105] have oft been recognized;[106] the standard practice both behind and within the doctrine has been to allow seemingly appropriate or positioning biblical texts and the elements of a systematic agenda to determine its meaning and expression. That is hardly surprising. It is to be expected. The dilemma that the "Anthropocene" presents is the way in

---

101. Dawson, *Extreme Cities*, 7–8.

102. Jon, *Cities in the Anthropocene*; Tyszczuk, *Provisional Cities*.

103. Pearson, "Cities."

104. Pearson, "God's Continued Providence."

105. See Van den Brink's essay in this volume.

106. Fergusson, *Providence*, 13–19.

which humankind (irrespective of the histories of colonialism, imperialism, extraction, racism)—the *anthropos*—has become a force of nature: the doctrine now finds itself face to face with those who claim this species has become, in effect, the creators of a new era, "earthmasters,"[107] and the mediators of extinction. The charge of *hubris* does not do away with the level of damage to the planetary system nor halt the slide into a cascade of tipping points. This doctrine that had once held out the hope of beneficent purpose, care, and meaning is rapidly spiraling off into issues of disbelief, theodicy, pastoral care in contexts of disaster management, and a practical response that lies well beyond the confines of theology. For some, the predicament identified by Thomas, Williams, and Zalasiewicz is really a "superwicked" problem.[108] The doctrine of providence now dwells in what is increasingly being described as a "planet on steroids."[109] It does so in a context where concerned nations are more likely to surround talk of climate with rhetoric to do with security,[110] climate engineering, and the technoverse. On matters to do with the "Anthropocene," climate change, and extreme weather events, the Christian faith is not the leading field of knowledge in responding to the evidence and likely impact of this deepening crisis. Its place is modest.

The increasing number of serious instances of angry weather nevertheless impose a sense of urgency upon talk of providence and provoke questions to do with audience. It is arguably the case that the event-based nature of climate attribution science fastens attention on particular times and places while assuming the universal and the global. The doctrine of providence presumes a public claim to do with confessions concerning God's foresight and care. Is it sufficient, then, for talk of this line of belief that is so often left implicit and inchoate to be little more than an in-house conversation among biblical scholars and theologians? Is it time for the church (and theology) to make more nuanced statements on climate-derived issues that confront specific places? Far too often, statements and marches gather a group of dissimilar causes together under one umbrella. That has often been the default practice in Australia, despite what is required in response to very different kinds of extreme events.

How an interpretation of the doctrine of providence (and other aspects of belief and praxis) engages with the predicament as it is encountered in cities, the bush, and low-lying islands is not the same. The researchers into climate change attribution have shown how important it is to engage with

---

107. Hamilton, *Earthmasters*.

108. See Levin et al., "Overcoming the Tragedy."

109. Gergis, *The Sunburnt Country*, 182.

110. Busby, *States and Nature*.

particular sites and do so on the grounds of prediction and consequence. What have theology, Scripture, and a Christian ethic to offer? Is it possible to craft "maps" that weave together the history and future of climate change in an actual site with a relevant representation of a Christian theology, in general, and providence, in particular? Is that not what is happening in this volume?

Writing on a polyphonic approach, Fergusson made a case for not binding providence only to the first article of the Christian creed. The very form of the elements of *conservatio*, *gubernatio*, and *concursus* imagine a God whose works are not confined to an act of creation and a subsequent sustaining of that act through the vagaries of life. It cannot be so "anchored" and be left separated from the Christological and Pneumatological.[111] It is a "dispersed" theology that "covers the full range of Christian doctrines"[112]—which makes the capacity for its use in the "Anthropocene" all the more pressing. The difficulties of discerning "what God is up to" in the unfolding suffering of creatures should not be allowed to let providence be seen as only being worthy of "antiquarian interest."[113] It must always possess a "promissory" character.[114] The witness of the past further warns against the lack of critical attention insofar as distortions can lead to forms of providentialism that should not be justified. The doctrine must be steered clear of attempts to pursue "private, factional or national self-interest."[115] In a time of deepening crisis, providence has a role to play in warding off that ever-present temptation. Its dispersed nature also implies that there is no need to seek to resolve the problems of other areas of the theological agenda upon which it may have some bearing—as is the case with theodicy.[116]

In this new era of the "Anthropocene," the notion of providence testifies to the overriding purpose of God; it does so despite the precariousness of life now that the *anthropos* finds itself having to dwell in the "afterlives" of the Holocene. Providence does not do away with the reality of evil and sin: they remain and, in some respects, are potentially intensified in the "Anthropocene." There have already been five mass extinctions of life: Elizabeth Kolbert warns of a sixth in the making,[117] and yet any reading of a

---

111. Fergusson, *Providence*, 58.

112. Fergusson, *Providence*, 297.

113. Fergusson, *Providence*, 302.

114. Fergusson, *Providence*, 305.

115. Fergusson, *Providence*, 299.

116. Fergusson, *Providence*, 303.

117. Kolbert, *Sixth Extinction*.

divine *conservatio* needs to acknowledge the continuing existence of creaturely life, even on a fast-changing planet. This world need not have existed at all: it could already have ended, but it has not. It still endures.[118] Under the rubric of the first article of the Christian creed, God has established "a providential work (of creation and preservation) within which God must subsequently work".[119] That is not undone by human agency throughout the great acceleration.

The question of "what is God up to" in this unfolding future without analogue is more likely to engage with the elements of *concursus* and *gubernatio*. It is tempting to suggest that policies of mitigation, adaptation, and resilience might represent the *anthropos* working, albeit unknowingly, in concert with the divine *concursus*. Were that to be so, then some caution is called for on at least a couple of grounds. The first has to do with correspondence—that is, with the risk of collapsing the remedial policies and practical responses of an endangered species into the divine *concursus*. From the perspective of a philosophy of the "Anthropocene," Hamilton has argued the need for an appropriate anthropology in this new era.[120] Is it enough to invoke ideas of stewardship, custodianship, and interdependence with the rest of creaturely existence without taking into account the power of the *anthropos* as a geologic actor and anthropogenic climate change? Is the *anthropos* the equivalent of what it meant to be human in the Holocene? The ever-present risk before a theory of providence in the "Anthropocene" is that it might imagine a beneficent role to be played out by humanity in a linear unfolding of history.

One way to minimize this possibility is to follow Fergusson in his turn to the hiddenness of God. Here he draws upon Luther and likens a theology of the cross (that also embraces incarnation and resurrection) to the *cantus firmus* of the doctrine. The dispersed nature of providence cannot ignore the way in which the promissory care and foresight of God are manifested in this most strange of ways.[121] The second article of the creed to do with Christ cannot be undone. It is likewise a marker of how God has put into place a mode of *gubernatio* that is more than setting the creation in place but is also redemptive in intent and consummation. Is there here a role for the Spirit in the "Anthropocene"? In the biblical traditions, it is energies and personal; it blows where it wills and is prophetic. It is concealed in sighs too deep for words, searches our hearts, intercedes in prayer, and binds the present to the future. It is wisdom and seeks out good news and freedom

---

118. Fergusson, *Providence*, 307.

119. Fergusson, *Providence*, 308.

120. Hamilton, *Defiant Earth*, 88-91.

121. Fergusson, *Providence*, 308-16.

and is not unfamiliar with the plight of suffering while waiting upon "the acceptable year of the Lord."

These theological thoughts may seem a far cry from Otto's angry weather and the work of event attribution science. That may be so, but it is likely that questions of providence will abound with increasing floods, fires, droughts, and the loss of biodiversity. They already lie hidden behind the experience of those who live on low-lying islands that are destined to become submerged in the not-too-distant future. The word providence (and its three elements bearing Latin names) may not be invoked, but it is assumed, often in a rough and ready way. It represents as such a protest against fate and luck. It signifies a hope that lies beyond the trajectory of a hothouse Earth that extends beyond its planetary boundaries.

The theological tradition that comes to us from the Holocene may have been hesitant to be too bold in its discourse on the specifics of a particular providence. That may well be the context for the notion in the "Anthropocene" where "the hiddenness of what is God up to" is more apparent in this predicament. There is perhaps an advantage in a rendering of a dispersed understanding of providence that is no longer simply seen as a sequel to creation: the inclusion of a theology of the cross and the activity of the Spirit invite a way of living in times that are likely to be ugly, a way of living which is derived from a call to compassion, grace, love of neighbor, hospitality, and justice to those most disadvantaged. Might this be part of the response to the question of where the providential work of the hidden God is revealed in the "Anthropocene"?

■ Bibliography

Allen, Miles. "Liability for Climate Change." *Nature* 421 (2003) 891–92.

Angus, Ian. *Facing the Anthropocene: Fossil Capitalism and the Crisis of the Earth System*. New York: Monthly Review Press, 2016.

Asafu-Adjaye, John, et al. "An Ecomodernist Manifesto." 2015. http://www.ecomodernism.org.

Benner, Suzanne, et al. *Paul J. Crutzen and the Anthropocene.* Berlin: Springer Nature, 2021.

Bergmann, Sigurd. *Weather, Religion and Climate Change.* New York: Routledge, 2020.

Breward, Ian. *Australia: The Most Godless Place Under Heaven?* Melbourne: Beacon Hill, 1988.

Busby, Joshua W. *States and Nature: The Effects of Climate Change on Security*. Cambridge: Cambridge University Press, 2022.

Chakraparty, Dipesh. *The Climate of History in a Planetary Age*. Chicago: University of Chicago Press, 2021.

Dalby, Simon. "Framing the Anthropocene: The Good, the Bad and the Ugly." *The Anthropocene Review* 3:1 (2016) 33–51.

Davis, Heather, and Etienne Turpin. *Art in the Anthropocene: Encounters Among Aesthetics, Politics, Environments and Epistemologies*. London: Open Humanities Press, 2015.

Dawson, Ashley. *Extreme Cities: The Peril and Promise of Urban Life in the Age of Climate Change*. Eugene: Pickwick, 2016.

Ellis, Erle C. *Anthropocene: A Very Short Introduction*. Oxford: Oxford University Press, 2018.

Fergusson, David. *The Providence of God: A Polyphonic Approach*. Cambridge: Cambridge University Press, 2018.

Funk, Chris. *Drought, Flood, Fire: How Climate Change Contributes to Catastrophes*. Cambridge: Cambridge University Press, 2021.

Gergis, Joëlle. *Sunburnt Country: The History and Future of Climate Change in Australia*. Carlton: Melbourne University Press, 2018.

Ghosh, Amitav. *The Great Derangement: Climate Change and the Unthinkable*. Chicago: Chicago University Press, 2016.

Gutiérrez, Julian E. "Who Is This King of Glory? Recovering the Identity of the God of Providence." In *Divine Action and Providence*, edited by Oliver D. Crisp and Fred Sanders, 55–73. Grand Rapids: Zondervan Academic, 2019.

Hamilton, Clive. *Defiant Earth: The Fate of Humans in the Anthropocene*. Crow's Nest: Allen and Unwin, 2017.

———. *Earthmasters: The Dawn of the Age of Climate Engineering*. New Haven: Yale University Press, 2013.

———. *Requiem for a Species: Why We Resist the Truth About Climate Change*. New York: Routledge, 2010.

———. "The Theodicy of the 'Good Anthropocene'." *Environmental Humanities* 7:1 (2016) 233–38.

Hamilton, Clive, Christophe Bonneuil, and François Gemenne, eds. *The Anthropocene and the Global Environmental Crisis: Rethinking Modernity in a New Epoch*. London: Routledge, 2015.

Helmer, Christine. "Providence: A Deflationary Mandate." In *Divine Action and Providence*, edited by Oliver D. Crisp and Fred Sanders, 74–95. Grand Rapids: Zondervan Academic, 2019.

Hu, Jane C. "The Decade of Attribution Science." *Slate*, 19 December, 2019. https://slate.com/technology/2019/12/attribution-science-field-explosion-2010s-climate-change.html.

Hughes, Lesley. "Climate Change and Our Life Support System." The Royal Society of New South Wales, 20 June, 2019. https://www.royalsoc.org.au/blog/women-science-lecture-3.

———. "Why Environmental Conservation Won't Save Nature." TEDxSydney, 26 November, 2021. https://tedxsydney.com/talk/why-environmental-conservation-wont-save-nature-lesley-hughes/.

Hulme, Mike. *Climate Change Is Not Everything*. Cambridge: Polity Press, 2023.

Islamic Declaration on Global Climate Change. 18 August 2015. https://www.ifees.org.uk/wp-content/uploads/2020/01/climate_declarationmmwb.pdf.

Jilly, India. "National Climate Emergency Rally, Melbourne, June 2009." *Greenlivingpedia.org*. https://greenlivingpedia.org/National_Climate_Emergency_Rally_Melbourne_June_2009/.

Jon, Ihnji. *Cities in the Anthropocene: New Ecology and Urban Politics*. London: Pluto Press, 2021.

Kolbert, Elizabeth. *The Sixth Extinction: An Unnatural History*. New York: Holt Paperbacks, 2014.

Kress, W. John, and Jeffrey K. Sine, eds. *Living in the Anthropocene: Earth in the Age of Humans*. Washington, DC: Smithsonian Books, 2017.

Lenton, Tim. *Earth System Science: A Very Short Introduction*. Oxford: Oxford University Press, 2016.

Levin, Kelly, et al. "Overcoming the Tragedy of Super Wicked Problems: Constraining Our Future Selves to Ameliorate Global Climate Change." *Policy Sciences* 45 (2012) 123–52.

McGuire, Bill. *Hothouse Earth: An Inhabitant's Guide*. London: Icon, 2022.

McNeill, J. R., and Peter Engelke. *The Great Acceleration: An Environmental History of the Anthropocene Since 1945.* Cambridge: Belknap, 2014.

Mendoza, S. Lily, and George Zachariah, eds. *Decolonizing Ecotheology: Subaltern and Indigenous Challenges.* Eugene: Pickwick, 2023.

Moses, Asher. "'Collapse of Civilisation is the Most Likely Outcome'. Top Climate Scientists." *Resilience,* 8 June 2020. https://www.resilience.org/stories/2020-06-08/collapse-of-civilisation-is-the-most-likely-outcome-top-climate-scientists/.

Moss, Jeremy. *Carbon Justice: The Scandal of Australia's Biggest Contribution to Climate Change.* Sydney: UNSW Press, 2022.

Muir, Cameron, et al. *Living with the Anthropocene: Love, Loss and Hope in the Face of Environmental Crisis.* Sydney: New South Publishing, 2020.

Neiman, Susan. *Evil in Modern Thought: An Alternative History of Philosophy.* Princeton: Princeton University Press, revised edition, 2015.

Newby, Jonica. "At Age Eight, Daniel Couldn't Swallow Food: How Climate Disasters Take a Toll." *The Sydney Morning Herald,* 17 June, 2023. https://www.smh.com.au/national/at-age-eight-daniel-couldn-t-swallow-food-how-climate-disasters-take-a-toll-20230505-p5d648.html.

Olsson, Lennart, et al. "Ethics of a Probabilistic Extreme Event Attribution in Climate Change: A Critique." *Earth's Future* 10:3 (2022) e2021EF002258.

Otto, Friederike. *Angry Weather: Heat Waves, Floods, Storms and the New Science of Climate Change.* Vancouver: Greystone Books, 2020.

Paul, Erik. *Australia in the Anthropocene: War Against China.* Singapore: Palgrave Macmillan, 2023.

Pearson, Clive. "Cities and the Challenge of Climate Change: Imagining 'Good Cities' in a Time of Dystopia." In *The Routledge Handbook of Religion and Cities,* edited by Katie Day and Elise M. Edwards, 426–38. London: Routledge, 2021.

———. "God's Continued Providence." In *T&T Clark Handbook of Christian Theology and Climate Change,* edited by Ernst M. Conradie and Hilda P. Koster, 395–405. London: T&T Clark, 2020.

———. "Is it Time for a Theological Step-Change?" In *Theology of a Defiant Earth; Seeking Hope on a Defiant Earth,* edited by Jonathan Cole and Peter Walker, 41–58. Lanham: Lexington Books, 2023.

———. "Unwrapping Theodicy." In *Theologies from the Pacific,* edited by Jione Havea, 181–92. Cham: Palgrave Macmillan, 2021.

Pearson, Jasmine, et al. "Climate-Driven Losses to Knowledge Systems and Cultural Heritage: A Literature Review Exploring the Impact on Indigenous and Local Cultures." *The Anthropocene Review* 10:2 (2023) 343–66.

Peterson, Thomas C., et al. "Explaining Extreme Events of 2011 from a Climate Perspective." *Bulletin of the American Meteorological Society* 93:7 (2012) 1041–67.

Ranawana, A. M. *A Liberation for the Earth: Climate, Race, and Cross.* London: SCM Press, 2022.

Ripple, William J., et al. "The 2023 State of the Climate Report: Entering Uncharted Territory." *BioScience* 73:12 (2023) 841–50.

Rothe, Delf. "Governing the End Times? Planet Politics and the Secular Eschatology of the Anthropocene." *Millennium—Journal of International Studies* 48:2 (2020) 143–64.

Shue, Henry. *Climate Justice: Vulnerability and Protection.* Oxford: Oxford University Press, 2014.

Skrimshire, Stefan. "Eschatology." In *Systematic Theology and Climate Change: Ecumenical Perspectives,* edited by Michael S. Northcott and Peter M. Scott, 157–74. New York: Routledge, 2014.

Sneed, Annie. "Yes, Some Extreme Weather Can Be Blamed on Climate Change." *Scientific American*, 2 January, 2017. https://www.scientificamerican.com/article/yes-some-extreme-weather-can-be-blamed-on-climate-change/.

Steffen, Will. "The Anthropocene: Challenges of the Human Age." Affinity Intercultural Foundation, 10 September, 2018. https://www.youtube.com/watch?v=RitK73xnB0M.

Steffen, Will, et al. "Trajectories of the Earth System in the Anthropocene." *Proceedings of the National Academy of Sciences of the United States of America* 115:33 (2018) 8252–59. https://doi.org/10.1073/pnas.1810141115

Talia, Maina. "Am I Not Your *Tu'akoi*—Your Neighbour? A Tuvaluan Plea for Survival in a Time of Climate Emergency." PhD thesis, Charles Sturt University, 2022.

Tasoff, Harrison. "Drought, Flood, Fire." *The Current*, 19 August 2021. https://news.ucsb.edu/2021/020377/drought-flood-fire.

Thomas, Julia Adeney, et al. *The Anthropocene: A Multidisciplinary Approach*. Cambridge: Polity Press, 2020.

Thuesen, Peter J. *Tornado God: American Religion and Violent Weather*. New York: Oxford University Press, 2020.

Trexler Adam. *Anthropocene Fictions: The Novel in a Time of Climate Change*. Charlottesville: University of Virginia Press, 2015.

Turner, Simon et al. "What the Anthropocene's Critics Overlook—And Why It Really Should Be a New Geological Epoch." *The Conversation*, 13 March, 2024. https://phys.org/news/2024-03-viewpoint-anthropocene-critics-overlook-geological.html#google_vignette.

Tyszczuk, Renata. *Provisional Cities: Cautionary Tales for the Anthropocene*. New York: Routledge, 2018.

Ward, M., et al. "Impact of 2019–2020 Mega-Fres on Australian Fauna Habitat." *Nature, Ecology and Evolution* 4 (2020) 1321–26.

White, Vernon. *Purpose and Providence: Taking Soundings in Western Thought, Literature and Theology*. London: T & T Clark, 2015.

Witze, Alexandra. "It's Final: The Anthropocene Is Not an Epoch, Despite Protest Over Vote: Governing Body Upholds Earlier Decision by Geoscientists amid Drama." *Nature*, 20 March 2024. https://doi.org/10.1038/d41586-024-00868-1

Yusoff, Kathryn. *A Billion Black Anthropocenes or None*. Minneapolis: University of Michigan Press, 2018.

Zalasiewicz, Jan, et al. "The Anthropocene within the Geological Time Scale: A Response to Fundamental Questions." *Episodes* 47:1 (2024) 65–83.

Zinke, Jens, et al. "North Flinders Reef (Coral Sea Australia) *Porites*.cp Corals as Candidate Global Boundary Stratotype Section and Point for the Anthropocene Series." *The Anthropocene Review* 23:1 (2023) 201–24.

On Providence, Suffering, and Creation Faith: Perspectives from "Anthropocene" Aporias

Elizabeth M. Pyne[1]

■ Approaching the Question

This volume is oriented by a call to explore narrative pathways that pass through the doctrine of providence: how and where can we find ways to tell the story of a God whose benevolent care for the created world is real, trustworthy, and, at least to some degree, evident amid the "Anthropocene," with its chaotic scrambling of known quantities and its pervasive tone of anxiety, of fear? How can we tell this story in ways that are in continuity with core claims of the Christian tradition while also reckoning with the potential of qualitatively different configurations of history, agency, power,

---

1. Elizabeth Pyne is assistant professor of religious studies and director of the William C. Sennett Institute for Mercy and Catholic Studies, Mercyhurst University.

> **How to cite:** Pyne, EM 2024, 'On Providence, Suffering, and Creation Faith: Perspectives from "Anthropocene" Aporias', in EM Conradie & UL Vaai (eds.), *Making Room for the Story to Continue?*, An Earthed Faith: Telling the Story amid the "Anthropocene", vol. 4, AOSIS Books, Cape Town, pp. 149-179. https://doi.org/10.4102/aosis.2024.BK415.07

and scale from those applicable to previous interpretations of this doctrine? That this call is framed in terms of "making room for the story to continue" gently alludes to the great challenge at hand: namely, the possibility that the story would have no such room, that this article of faith has run its course and, now rendered moot or meaningless by present planetary conditions, would fall by the wayside as a chapter of centuries past.

Truth be told, an interest in providence as such has not been atop my list of theological priorities. My training lies largely, although by no means exclusively, in the Catholic tradition. Having been shaped by feminist and liberationist perspectives as well as sustained engagement with varieties of critical theory and continental philosophy, I have tended to regard this element of Christian faith with a blend of distance and distaste. Too often, claims about divine providence reverberate between abstract attributes (e.g. omnipotence; omniscience), resulting in propositional truth claims that are less than edifying in the face of evil, anger, and grief.[2] These theodical endeavors may also, as the editors' introduction aptly notes, cause further harm by sanctioning or legitimating the suffering of others or one's own.[3] On this point I want to insist, with certain political and liberation theologians, that there can be no *theoretical* reconciliation of the problem of suffering and evil.[4]

As an exponent of liberation theology writing from the Global North, the Flemish Dominican Edward Schillebeeckx has been especially influential to me in shaping this conviction.[5] He writes in no uncertain terms that "God does not want human beings to suffer but wills to *overcome* suffering where it occurs in our history."[6] Yet simultaneously, his later work brings to the forefront an awareness of the scope and scale of suffering in historical experience; indeed, the enormity of human suffering becomes "the *locus theologicus* of Schillebeeckx's theological enterprise," as one commentator puts it.[7] And so the main lines of an impasse are drawn: to give a determinate place to suffering in any story of divine purpose or plan is to gainsay what Schillebeeckx sees as the good news of salvation; meanwhile, the persistence of suffering would seem to refute any rationally consistent

---

2. See Tilley, *The Evils of Theodicy*.

3. See also Clive Pearson's summary of Christine Helmer's "deflationary argument" regarding the doctrine of providence in his essay on "Angry Weather" in this volume.

4. In addition to Edward Schillebeeckx as an example (below), see Elizabeth Johnson's discussion of "the failure of theodicy" in *Quest for the Living God*, 51-60.

5. See Lee, "Schillebeeckx and the Path to a Liberation Theology."

6. Schillebeeckx, *Christ*, 723 (original emphasis). For Schillebeeckx, God is "pure positivity." See Schillebeeckx, *God Is New Each Moment*, 10.

7. Rego, *Suffering and Salvation*, 10.

account of the saving God's activity in history.[8] A negative check on rational meaning therefore appears as both a spiritual and ethical prerogative: "The Christian message does not give an explanation of evil or our history of suffering. [...] Even for Christians, suffering remains impenetrable and incomprehensible, and provokes rebellion."[9]

Yet neither is silent withdrawal from the scene a compelling option. For this question about divine care arises from real anguish, which demands some attempt to speak coherently amid personal and historical tragedy about the living God who is Love.[10] When the meaning of the doctrine of providence is relocated from quasi-mathematical calculations among lofty predicates to its core of "trust in God's loving care," it becomes clearer how this notion is of non-negotiable importance. In preparing this essay, I have come to see providence as a kind of connective tissue between understandings of creation and salvation—understandings that are crucial in my work and in some of the most pressing problems taken up by feminist, liberationist, and ecological projects. In particular, the *An Earthed Faith* project has prompted me to explore the resonance between providence and Schillebeeckx's notion of "creation faith."[11]

■ Guiding Insights: Schillebeeckx on "Creation Faith"

Schillebeeckx expressed this theme throughout his work as an extension of the characteristically Catholic principle, drawn from Aquinas, that "grace builds on nature."[12] He came to emphasize it as a discrete and especially significant reference point in his later work.[13] At its heart is the conviction that salvation should never be understood as salvation *from* creation, but

---

8. Mary Catherine Hilkert summarizes: "He began to see that 'scarlet thread' of radical and senseless suffering on a global scale as the greatest challenge to belief in a loving Creator God with the power to save humankind." In "The Story of Jesus and Human Flourishing," 279.

9. Schillebeeckx, *Christ*, 721.

10. For mid-twentieth-century theologians persuaded of the failure of theodicy, bourgeois indifference to suffering was also a pressing problem to be addressed; see Johnson, *Quest for the Living God*, 55.

11. Translated from the Dutch *scheppingsgeloof*. The term sometimes appears in hyphenated form as "creation-faith."

12. As noted by Hilkert, "The Story of Jesus and Human Flourishing," 269.

13. See, e.g., Schillebeeckx, *I am a Happy Theologian*, where he comments, "I regard the creation as the foundation of all theology" (47). Schillebeeckx's corpus is variously parsed in terms of early and later work, with a significant shift around 1966-1967, or a tripartite scheme along the lines of that proposed by William Portier: a movement from "the interpersonal categories of phenomenology (1957-66) to hermeneutics (1966-71) to critical theory (1971 to the present)"; see Portier, "Interpretation and Method," 26 (written before Schillebeeckx's death in 2009).

instead as its fulfillment. This allows for a strong affirmation of the fundamental goodness of creation in its finitude; sin, rather than finitude, is the enemy. Another corollary is the premium Schillebeeckx places on human freedom within history as the mediation of God's graced activity in the world.[14] In the middle phase of his writing, this principle took shape around the priority of the future. Finite freedom is a gift and a unique, unsubstitutable task: human beings are called to help shape a better future, and Christians should see these efforts as fragmentary instantiations of God's promise to create "a new heaven and a new earth."[15] Yet Schillebeeckx is consistently and acutely aware of the contingency and limitation of created finitude in its autonomy and the potential for sinful enactments of freedom. He sees a relation between the "absolute presence of the creating God"—which is equally the "absolute, saving presence of the living God"—and the world that is not-God.[16] Thus the "faith" in question is neither about a set of propositions that explain the physical world, nor a starting point on a linear chronology, nor some kind of "mid-point of a continuum between dualism, at one extreme, and pantheism at the other."[17] Creation faith holds together God's transcendence and immanence as "mutually co-constitutive," Martin Poulsom observes, in a "dialectic […] in which they exist in direct proportion to each other, not inverse proportion."[18] Likewise, it should by now be clear that Schillebeeckx's references to creation faith are never to creation alone—the work of some distant God—but instead treat the "symbol" of creation as part of the story of who God is and what God's love does within the finite realm of nature, history, and human life:

> Creation is an act of God which, on the one hand, unconditionally gives us our own particular character—finite, not divine, and destined for true humanity. Simultaneously, on the other hand, creation is an act in which God presents himself in selfless love as our God: our salvation and happiness, the supreme content of what it means to be true and good humanity. God creates men and women freely for the salvation and happiness of man himself; however, in the very same act, with the same sovereign freedom, he wills himself to be the deepest meaning of human life, its salvation and happiness.[19]

Schillebeeckx does not tend to use the term "providence" when describing this relation, but his comments point in a similar direction when he writes:

---

14. See Hilkert, "The Story of Jesus and Human Flourishing," 272.

15. See Schillebeeckx's homily on this image from the Book of Revelation in the collection *God Among Us*, 144–48. The phrase features frequently in his writing elsewhere. See also Claesson, "Edward Schillebeeckx."

16. Compare two parallel passages in which these respective phrasings appear: Schillebeeckx, *Interim Report*, 106 and *Church*, 231.

17. Poulsom, "Schillebeeckx's Praxis of Creation," 242.

18. Poulsom, "Schillebeeckx's Praxis of Creation," 243.

19. Schillebeeckx, *Interim Report*, 114.

"[G]od is 'with us' in all that this finitude entails, both in positive experiences and in failure, suffering, and death," precisely because God "is after all the creator of this entire 'saeculum', so that there are no eras, no centuries and not even any hours in which he leaves no witness to himself."[20] Once again, though, that divine presence is not static and, relatedly, it does not negate the contingency and freedom of human beings as those to whom "God has entrusted creation [...] not merely as caretakers of a past condition, but as co-creators with God of the future."[21] Refracting the Thomistic approach to nature and grace in this new historical key allows Schillebeeckx to pursue the imperative of communicating religious wisdom in an increasingly secular world. Roger Haight glosses the Dominican's exposition of creation faith in terms of a "basic trust": "[t]he possibility of speaking of a creator God is founded in a 'basic trust' experienced within a person and directed toward the human project itself." "The Christian interpretation of this secular trust names it a hope for salvation," Haight continues, echoing Schillebeeckx's insistence that this is not a *proof* of God's existence or designs.[22] For some Christian commentators, Schillebeeckx's merging of trust in God's care with a humanistic rendering of existential trust would represent a departure from traditional notions of providence. Yet, precisely because of his understanding of creation (as the relation between God and the world outlined above), this dialogue with the secular, and the associated movement between religious and ordinary language, was never intended to be a mere concession to contemporary demands. Instead, it shows forth the heart of the gospel as good news: God's cause is the human cause, Schillebeeckx avowed, such that there can be "no salvation outside the world."[23] In connecting creation and salvation as dimensions of the work of the same loving God, Schillebeeckx's elaboration of creation faith seems to occupy a structural position akin to that given to providence in the editors' introduction, showing particular investments in the classic themes of *creatio continua* and *concursus*.[24]

---

20. Schillebeeckx, *Interim Report*, 105 (cited by Poulsom, "Schillebeeckx's Praxis of Creation," 246). Poulsom notes that (as above) "creating God" is a better translation of Schillebeeckx's Dutch "*de scheppende God*" than is "creator God."

21. Haight, "Christ and Culture," 399.

22. Haight, "Christ and Culture," 397.

23. For two examples of these recurring phrases, see Schillebeeckx, *Interim Report*, 113 and *Church*, 12.

24. See Conradie and Vaai, "In God We Trust?" The resonance is with their claim that providence is an indispensable connecting element, although not one that can stand alone from other systematic categories. However, I recognize a greater willingness on Schillebeeckx's part always to incorporate God's work of salvation.

Creation faith remained a core theme in Schillebeeckx's later writings on Christology and soteriology, the work for which he is best known.[25] The elements we have seen thus far help to illuminate this "later soteriological anthropology," the curvature of which was increasingly shaped around the emphasis on suffering noted in my initial discussion of theodicy. Creation faith is the ontological frame for the claim that God does not want human beings to suffer, a truth poignantly and passionately embodied in the "story of Jesus," the savior whom Schillebeeckx understands as "concentrated creation."[26] At the same time, this framing also underscores the challenge that suffering poses to faith in a providential God, if we continue here to track this idea of providence as the connective tissue between the experience of creation and salvation. Mary Catherine Hilkert aptly expresses how the dilemma takes shape in Schillebeeckx's survey of the defining features of the world in the last quarter of the twentieth century: "Schillebeeckx identified the multiple and pervasive threats to human well-being, systematic forms of injustice, and challenges to hope around the globe as evidence of the 'absence of salvation'. That absence took on two distinct but interrelated forms: on the one hand, the dehumanizing and meaningless suffering of two-thirds of the world's population and, on the other hand, the lack of evidence of solidarity among the other one-third, who were focused on preserving power and their own standard of living and who viewed 'the other' as a threat."[27]

In the relational model of finite creation held in the absolute presence of God, challenges to the ground of human hope and ethics become the other side of the coin to theodicy. It is true that Schillebeeckx writes of the "defenceless superior power of God," a vulnerability that represents a kind of "divine yielding" to creation's autonomy.[28] However, this should not be interpreted as a tidy separation of human freedom from divine power; it is neither a denial of God's immanence nor—worse yet—a mark of God's indifference.[29] Recall: God's cause is the human cause, and the cause of all creation. Ultimately, Schillebeeckx's theology disallows passing the buck

---

25. Hilkert, "The Story of Jesus and Human Flourishing," 279. This period coincides with what Kathleen McManus refers to as Schillebeeckx's "political awakening"; see McManus, *Unbroken Communion*, 67.

26. It is vital to note the thorough integration of Christology and creation faith in Schillebeeckx's understanding. Only in the interest of space do I leave "the story of Jesus" somewhat to one side. For one recent study of his evocative characterization of Jesus, see Lewis, *Concentrated Creation*.

27. Hilkert, "The Story of Jesus and Human Flourishing," 283.

28. See Schillebeeckx, *Church*, 83–89, and his essay "Doubt in God's Omnipotence: 'When Bad Things Happen to Good People'," in *For the Sake of the Gospel*, 88–102. Here and elsewhere, he shows considerable interest in the idea of God's hiddenness and the Christian apophatic tradition, mentioned by several other contributors in this volume.

29. See Kennedy, *Schillebeeckx*, 92.

from God to leave the problem of suffering (un)comfortably with human beings: anthropodicy is as much a theological stumbling block.[30]

Creation faith is therefore resolutely not a solution to the problem of creaturely suffering and will militate against any views that take divine providence to provide such a solution. Yet it can nonetheless give some direction to the idea of "reconciling" that spurs this volume's inquiries. For Schillebeeckx, creation faith has a "critical and productive force."[31] The notion of "basic trust" identified earlier dislodges both "overly pessimistic and optimistic conceptions of human history and society" and simultaneously "translates through freedom into commitment to this world and history."[32] Theology in the "narrative-practical key" adopted by Schillebeeckx in the latter part of his career would emphasize that the telling of the Christian story must be bound up with praxis, that is, "the dialectical interacting of theory and concrete action."[33] Reflecting on the life and death of Jesus of Nazareth in dialogue with critical theory shapes, for him, the truth of Marx's push not only to interpret the world but to change it.[34] Here, suffering plays a crucial role.

The transformative potential within creation faith is hinged to the critical, epistemic, and productive force of suffering as an experience of contrast, an idea that Schillebeeckx honed over the course of several decades. Negative contrast experiences carry an "implicit ethical demand" by showing up the distance between a "'no' to the world as it is" and the intuition of what he calls the *humanum*: a longing for happiness; a true and livable humanity; a promise of flourishing that lies ahead of us, unrealized. The experienced negativity of this "no" is a seedbed of indignation and, in turn, of protest. Schillebeeckx considers this contrastive mode a basic, "pre-religious" human experience.[35] As such, it is a point of dialogue between secular and religious interpretations of reality at a moment when "the quest for salvation," in one form or another, has become paramount. We have seen already how movement between these interpretations is necessitated by Schillebeeckx's view of God and the world, with its enduring conviction regarding nature and grace. The contrastive paradigm stresses

---

30. See Schillebeeckx, *Christ*, 713–14.

31. Schillebeeckx, e.g., *God among Us*, 102.

32. The quoted lines are from Kennedy, *Schillebeeckx*, 89 and Haight, "Christ and Culture," 398.

33. "Narrative-practical key" is Hilkert's shorthand for the methodological approach characteristic of Schillebeeckx's later work ("The Story of Jesus and Human Flourishing," 279) while the definition of praxis is found in Hill, "A Theology in Transition," 3.

34. Schillebeeckx, *The Understanding of Faith*, 24.

35. For one of several expositions of this point, see Schillebeeckx, *Church*, 5. See also Hilkert's "Experience and Revelation."

yet more strongly that the saving God is on the side of those who experience suffering, with them in their "no" to suffering and in their search for healing, hope, and peace.

God's providential care here is, as ever, mediated.[36] Schillebeeckx's interlinked understandings of creation faith, Jesus—who is "at once the parable of the living God at work in history and the paradigm of humanity fully alive"—and discipleship all feature in his account of how "the mystical praxis of Christianity is linked with its social and political praxis through the non-theorizable mediation of the history of human suffering."[37] In her study of this theme, Kathleen McManus writes that "the complex implications of suffering in his theology resolve into a praxis of solidarity."[38] For Schillebeeckx, "God's salvific intention for creation is realized whenever and wherever evil is resisted and good is furthered."[39] The simplicity of this formulation perhaps belies the difficulty of the work. All the more so, we might venture, in the context of "wicked moral problems" such as climate change, where lines between oppressor and oppressed become tangled to some degree.[40] While his own analyses of socio-historical causes of suffering tended to remain fairly abstract,[41] the core of this theological method retains immense value for contemporary reflection on providence. Schillebeeckx incorporated a robust critical apparatus into his narrative of solidaristic hope. Any enactment of creation faith—in the sense of a basic trust in a human creaturely future, of which God is the final guarantor—

---

36. O'Meara helpfully explains Schillebeeckx's position: "The structure of the presence of God as mediated immediacy implies that the human response to God has a similar structure of mediation. In other words, although religion cannot be reduced to cohumanity and sociopolitical concerns, it cannot do without this mediation. God's saving power never breaks in from outside human history. Rather, God's grace is present in the structure of historical human experience and praxis. Because of the constitutive relationship between personal identity and the social structures that provide freedom, sociopolitical improvements form an integral part of what is experienced as the grace of God"; "Salvation: Living Communion with God," 109.

37. For this description of Jesus, see Hilkert, "The Story of Jesus and Human Flourishing," 279 (glossing Schillebeeckx, *Jesus*, 626); the quote on praxis is from Schillebeeckx, *The Language of Faith*, 124.

38. McManus, *Unbroken Communion*, 164.

39. O'Meara, "Salvation: Living Communion with God," 100. One should note that this claim about salvation realized in history is inseparable from "the story of Jesus." O'Meara summarizes: "The surmounting of death in the resurrection is God's definitive acceptance, validation, and final fulfillment of the earthy life of Jesus. It is precisely in this way that the cross of Jesus acquires a productive and critical force in the dimension of human history. In spite of the disruption caused by the fatal rejection of Jesus' person and message, there is continuity between the hidden dimension of what took place on the cross and its manifestation in the resurrection, namely, the living and unbroken communion between Jesus and God" (112-13). That the resurrection is definitive does not mean, however, that it is the end of the story. Schillebeeckx asserts that this "force" becomes effective through Jesus's followers and all those who undertake the praxis of the reign of God. Their lives constitute, he says, a "fifth gospel"; see Schillebeeckx, *Christ*, 2.

40. Riley and Bauman, "Wicked Problems in a Warming World."

41. As observed by Hilkert, "The Story of Jesus and Human Flourishing," 283 n. 13.

must train itself on the criticism of suffering, not its justification. Crucially, given the topic at hand in this volume, Schillebeeckx maintains that such scrutiny should be aimed at any ideology or form of knowledge, be it scientific or religious, which conceals injustice and perpetuates needless suffering.[42]

This sketch of a major twentieth-century Catholic thinker will guide my subsequent and more expansive inquiry into considerations of providence in the "Anthropocene." Two main lessons are at the fore. Firstly, if reference to this aspect of the Christian story is to continue, there is a strong case to be made that it should be understood in experiential and relational terms.[43] It is "God-talk" about the one in whom human creatures place their trust. While its object is God, perhaps in a particular way providence also describes those who articulate it. For Schillebeeckx, God and creation are distinct but indelibly linked together. And, with respect to the mystery of suffering, it is finally we creatures who must do the reconciling. Thus, an inquiry into providence must address theological anthropology—the human condition—as much as any doctrine of God.[44] Consequently, thinking with Schillebeeckx on creation faith suggests that the notion of a geohistorical epoch centered on the *anthropos* is especially important territory for reflection on providence: indeed, those implications are not secondary questions but, on this model, are present within the ways that the "Anthropocene" reflects human identity and meaning.[45]

The second point of guidance returns us to the rejection of theodicy with a sharpened sense of the role of "critical negativity" in making room for the story to continue. As a reflexive hermeneutical endeavor, the theological process of reconciling reaches in some sense beyond human

---

42. Schillebeeckx characterizes the two main options in this way: "the purposive, emancipatory type of knowledge characteristic of the sciences and technology" and "the various types of contemplative, esthetic and playful, so-called 'aimless' knowledge, which keeps turning around its object." He continues: "[I]n our given human predicament and within the context of our concrete social culture, contemplation and action—paradoxically but nevertheless truly—can be brought together only by the criticism of the accumulated history of human suffering and the ethical consciousness developed in the course of this history of suffering." Schillebeeckx, *The Language of Faith*, 124-25.

43. See Gijsbert van den Brink's contribution in this volume for an enriching biblical commentary that complements this point.

44. In a relatively early essay, "Dialogue with God and Christian Secularity," Schillebeeckx makes the point directly: "[W]e cannot approach man's essence better than by doing so through reflection on divine Providence" because "the question as to the meaning of human life coincides with that regarding divine Providence." Commenting on this passage, Cooper reinforces a thread I have emphasized throughout this presentation of Schillebeeckx's work: "he describes the relationship between these two realities in a manner that does not negate the human or the divine terms of the relationship but maintains their distinctiveness"; see her *Humanity in the Mystery of God*, 4.

45. Conradie writes, "To assess the 'Anthropocene' entails nothing less than an assessment of the human condition." See "Some Theological Reflections on Multi-disciplinary Discourse on the 'Anthropocene'," 11.

experience toward the transcendent but remains—or must remain, I would argue—bound to honesty about the catastrophic undoing that suffering in history represents. If the parameters Schillebeeckx sets are sound, and I think they are, a theologian can only ever tell a story that holds in place this tension. I approach this volume's guiding question concerning "reconciliation" in this light, trying to avoid overly abstract or closed formulations in favor of seeking just enough narrative thread to go on. While the discursive analysis of "Anthropocene" framings that follows is at one remove from a direct exploration of what God is doing, Schillebeeckx's theology bridges the gap. Against the backdrop of creation faith, the interpretive process itself becomes in some sense a site of fidelity. The goal is something like what Donna Haraway calls "staying with the trouble."[46]

Amid the "Anthropocene"

How does the question posed differ from a version that asks more generally, "How could the suffering of God's creatures be reconciled with trust in God's loving care"? What difference does it make to treat creaturely suffering (and avoid theodicy) in the "so-called 'Anthropocene'"? I want to suggest that an attempt to make room for the Christian story *here* benefits from testing the limits of two oppositional framings: "everything has changed" and "nothing has changed." As Christophe Bonneuil and Jean-Baptiste Fressoz, among others, have argued, narrative structure surrounding the "Anthropocene" matters a great deal—both in itself and for the way in which it shapes responses.[47] So it does for theological inquiry and the crucial questions under discussion: What kinds of experiences are deemed to be suffering? Whose suffering counts? Which instances of suffering elicit protest and spur change, and which meet rationalizing dead-ends? Within these conceptions, how do human, nonhuman, and divine agencies interact? The goal of this dual exploration is not any sort of idealized middle position. Rather, my contention is that juxtaposing maximalist and minimalist interpretations of the contextual significance of "the Anthropocene" reveals, from each side, a set of tensions and instabilities that are instructive.

Everything Has Changed?

This first option takes as its point of departure an emphasis on the newness of the planetary conditions that obtain. These conditions are such that they warrant the designation of an epochal geological transition and indeed a

---

46. Haraway, *Staying with the Trouble*.

47. Bonneuil and Fressoz, *The Shock of the Anthropocene*.

transition in the Earth System: *anthropos-kainós* (latinized as *cænus* to give -cene). But while this new epoch is one marked in a sequence of "-cenes"— and thus demonstrates some form of continuity with what came before— for many, the salient claim is about novelty as stark discontinuity. In one of their early reflections on "the Anthropocene," Paul Crutzen and Will Steffen contend that "[t]he Earth currently operates in a state without previous analogy."[48] Beginning in 2001, the Intergovernmental Panel on Climate Change (IPCC) began to write of "tipping points," defined as "critical threshold[s] beyond which a system reorganises, often abruptly and/or irreversibly."[49] What is at issue, according to Johan Rockström and twenty-eight collaborators, are "planetary boundaries"; these are "dangerous thresholds" that signal the outer edges of a "safe operating space for humanity."[50] As of this writing, Rockström and others indicate that six have already been crossed.[51] While discussion around the "Anthropocene" still includes ample references to risk and future threats, one of its rhetorical effects is to focus attention on the present and on what has already transpired. One can witness this trend with regard to climate change in particular (though by no means exclusively here): in his essay in this volume, Clive Pearson considers the increasing confidence within "new climate science" to attribute extreme weather events to anthropogenic climate effects—practically in real time. Anecdotally, many of us may find the heartrending headlines that accompany such events to have become part of a dystopian "new normal." Scientifically, there is ample evidence that whatever ameliorative measures are pursued (and let us hope they are many more than presently undertaken) "there is," as Hannes Bergthaller and Eva Horn write, "no going back." For them, "[t]his is why the Anthropocene is more than just a crisis—it is a radical break: a break from the unusually stable ecological conditions that characterized the Holocene."[52]

The second facet of this claim for an unprecedented kind of planetary change hinges on the much-discussed *anthropos* named in this new epoch. The "Anthropocene" is often described as a fundamental shift in the relation between humanity and nature. As Simon Lewis and Mark Maslin summarize in the opening pages of *The Human Planet: How We Created the Anthropocene*, when considered against the course of the planet's 4.5 billion-year history, "now there is a new force of nature changing Earth:

---

48. Crutzen and Steffen, "How Long Have We Been in the Anthropocene Era?," 253.

49. IPCC, "Summary for Policymakers."

50. Rockström et al., "Planetary Boundaries."

51. Stockholm Resilience Centre, "Planetary Boundaries."

52. Horn and Bergthaller, *The Anthropocene*, 2.

homo sapiens"; whether on the order of decades, centuries, or millennia (their review of the dating debates results in an early seventeenth-century "Orbis Spike"), the Earth has become a "human-dominated planet."[53] Similarly, from Steffen and colleagues: "the human imprint on the global environment has now become so large and active that it rivals some of the great forces of Nature in its impact on the functioning of the Earth System."[54] For many, this is the key point: a new picture of causality on a geological and global scale. Human behavior is responsible for the transition described above. The weighty critique that it is in fact only some humans and specific socio-economic systems that bear responsibility will receive further treatment later in this essay. What is notable for the moment is the observation that such a critique does not necessarily dislodge the basic affirmation of a decisive, qualitative shift in planetary conditions. So, for instance, one of the key proponents of the argument that this global epoch is better understood as the "Capitalocene" also refers to it as a "no-analogue state."[55]

Although not all appraisals of this shift are negative or at least not wholly so (a point to which we will return in the next section), the predominant attitude conveyed in this narrative framing—"everything has changed"—sees a deeply worrying trend. References to damage, harm, injury, loss, destruction, disaster, catastrophe, and apocalypse are widespread. For our guiding question regarding creaturely suffering, then, the implications are potent. In the "Anthropocene" we are witnessing, first and foremost, a multiplication and magnification of suffering. Climate change has a demonstrable death toll. Floods and wildfires are among its most terrifying leading edges. A suite of deleterious effects on human health and well-being, from respiratory diseases to lack of access to education, follows from intersecting environmental crises of anthropogenic origin. Rob Nixon has written powerfully about these forms of "slow violence" and their disproportionate impacts on poor and vulnerable people across the globe.[56] Similarly, degradation and destruction of habitats, along with resource scarcity and the sometimes-violent conflicts spurred thereby, is forcing communities to migrate in vast numbers. Needless to say, these climate migrants frequently face political hostility leading to further threats to livelihood and life.

---

53. Lewis and Maslin, *The Human Planet*, 3.

54. Steffen et al., "The Anthropocene: Conceptual and Historical Perspectives," 842 (cited in Bonneuil and Fressoz, *The Shock of the Anthropocene*, 16).

55. Jason Moore, cited in Horn and Bergthaller, *The Anthropocene*, 2.

56. Nixon, *Slow Violence and the Environmentalism of the Poor*.

For many, what also comes into view—and in fact is sometimes in the foreground ahead of human suffering—is the suffering of (nonhuman) nature. This is part of the distressing uniqueness of this new epoch. If human beings have long had the capacity to alter local or regional environments and have previously engaged in activities with measurable impacts on flora and fauna, now such capacity is exponentially increased to touch global systems, with impacts which in turn manifest in cascading effects on various creatures. While there are deep debates about the experiential quality of nonhuman pain and death, it seems hard to deny the fittingness of speaking of suffering in regard to animals burned alive in the same wildfires that have taken human life or, more broadly, when they are denied the requisite components of their natural life cycle (e.g., food and habitat). By extension, one can point to a staggering loss of biodiversity, even constituting a "sixth extinction,"[57] and to the multifaceted problems linked to glacial melting, deforestation, and ocean acidification.

Insofar as "the Anthropocene" directs attention to the scope and scale of such impacts, it carries forward and then amplifies a central tenet of the environmental movement, namely that some human activity is profoundly damaging, so much so that it now occupies a qualitatively different register. That insight lies at the heart of the claim that something significant *has* changed for theological inquiry into suffering and providence. An expanded purview on suffering further problematizes the question of reconciling this reality with God's providential care. Although without explicit reference to "the Anthropocene," the vivid language of certain ecotheologians is illustrative here. In his prophetic book *Cry of the Earth, Cry of the Poor*, Leonardo Boff refers to "systematic assault on the Earth" as one of "two bleeding wounds."[58] More recently, Pope Francis has adopted Boff's perspective in his substantial engagement with religious environmentalism. In the opening passage of his 2015 encyclical *Laudato Si'*, he blends Boff's metaphorical language with Saint Francis of Assisi's personified imagery and with biblical texts on creation:

> This sister now cries out to us because of the harm we have inflicted on her by our irresponsible use and abuse of the goods with which God has endowed her. We have come to see ourselves as her lords and masters, entitled to plunder her at will. The violence present in our hearts, wounded by sin, is also reflected in the symptoms of sickness evident in the soil, in the water, in the air and in all forms of life. This is why the earth herself, burdened and laid waste, is among the most abandoned and maltreated of our poor; she "groans in travail" (*Rom* 8:22). We have forgotten that we ourselves are dust of the earth (cf. *Gen* 2:7); our very bodies are made up of her elements, we breathe her air and we receive life and refreshment from her waters.[59]

---

57. Kolbert, *The Sixth Extinction*.

58. Boff, *Cry of the Earth*, 104.

59. Pope Francis, *Laudato Si'*, 2.

While the Christian inflection adds distinctive notes, several constitutive elements of this diagnosis are largely consistent with secular evaluations of what is different about the "Anthropocene." Narratives of overbearing anthropocentrism—often tied to the "great divide" of culture/history from nature as a major signifier of Western modernity—are omnipresent in evaluations of ecological crises, climate change, and the "Anthropocene."[60] The "sin" language that Francis and other religious commentators make explicit is arguably implicit in a great deal of this environmentalist and scientific analysis.[61]

To the extent that something like a sin–consequence (sin–punishment?) paradigm seems to be applicable, this might be taken to suggest that the "Anthropocene" as context for theology can be accommodated fairly easily into a known narrative structure. Not so, say the maximalists. Notwithstanding questions about whether such territory around "the wages of sin" has ever been settled, such a confident assertion misses the mark with respect to the gravity, extent, and unpredictability of what the so-called "Anthropocene" names and seeks to reckon with. Attesting this shift brings in two profound existential questions. Will movement outside a "safe operating space for humanity" mean the extinction of the human species? Do these changes to the Earth's major systems spell the end of all or nearly all biotic life? Some say these are overblown fictions with no more value than cheap fearmongering; others maintain that they are precisely the questions that need to be confronted.[62] Much rhetoric around the "Anthropocene" activates a poignantly tragic register: because too many of us—perhaps especially many Christians—have refused the proposition that we belong here, that we are at home on the Earth,[63] too many of us have lived in ways that mean we may soon face an uninhabitable Earth.[64] A sentiment often intended to inspire meliorative responses, it can also be skewed the other way towards resignation. Sometimes a clear-eyed accounting of negative human impacts on nonhuman entities and the planet is tied to what Adam Kirsch has called "Anthropocene antihumanism"—a collection of views that would welcome or even facilitate the disappearance of *homo sapiens*.[65]

Any of the interpretive possibilities regarding such cataclysmic outcomes is a theological nightmare. Would God be spoken of as Creator

---

60. See, among others, Descola, *Beyond Nature and Culture*.

61. See Conradie, *Secular Discourse on Sin in the Anthropocene*.

62. See, e.g., Cohen et al., *Twilight of the Anthropocene Idols*; Hamilton, *Requiem for a Species*.

63. See Scott, "The Re-Homing of the Human."

64. Wallace-Wells, "The Uninhabitable Earth."

65. See Kirsch, *The Revolt Against Humanity*.

only in the past tense? This has never been the Christian conviction (see: *creatio continua*). Nor is it clear what salvation could mean; the temptation is to separate out God's saving work, which usually means extracting human beings from their creaturely status—falling back on a heavenly rather than earthly destiny and ensoulment rather than embodiment. Such maneuvers are, ironically, far too close to the "moonshot" thinking of the Elon Musk and Jeff Bezos crew. Absent spiritualizing or techno-elitist escape mechanisms, the "Anthropocene" thus constitutes an intensification of the problem of suffering that is at once overwhelming and acute. The editors' question in the introduction to this volume is apt, and then some: "Does it even make sense to say that God cares? Does this conviction not die the death of a thousand qualifications, by every experience of unjustifiable suffering?"[66]

Nothing Has Changed?

A version of the response that, in fact, the two questions do not differ at all could follow from a head-in-the-sand ignorance of any perceptible problem. One need not imagine such a response, as there are myriad examples of downplaying or denying climate change and a host of environmental challenges. Theologically, it is easy to see how such framings align with a fideistic perspective: nothing *can* change, really, because God is God and the world is the world. Context, in this view, is basically irrelevant for the practice of theology. Within these parameters, one might often observe a rather rosy confidence around God's providential power: nature chugs along as per God's design for it cannot do otherwise; and history more or less does too, notwithstanding what might appear—to the insufficiently believing—as examples to the contrary. Alternatively, consider a fatalistic variant: it has all been downhill since the fall, hence a leveling effect in regard to any supposedly new or different questions for a faith that waits for the divine clean-up act. These narratives are, in more or less obvious ways, deflationary with respect to the imperative to protest against suffering, to dare to say that at least some of the brokenness of the world and the harm we do to each other cannot be seen as consistent with God's will for creation. Moreover, even as they resolve the existence of suffering in a single sweep, such perspectives almost inevitably re-inscribe a dualistic picture of nature and history and of the causality that comes with this divide: one set of explanatory mechanisms obtains in the nonhuman world and another, related to free will, characterizes human activity. Both are trained on impacts affecting human beings; the possibility of suffering in the nonhuman natural world is obscured from view.

---

66. See Conradie and Vaai, "In God We Trust?" in this volume.

Suffice it to say I do not see this as a defensible position. However, there are ways of calling into question a narrative structure where the sheer newness of "the Anthropocene" as context for theology is assumed. Specifically, I want to pursue the claim that there are significant points of contact between twentieth-century developments around the mystery of suffering—as expressed in the work of one Catholic theologian—and what it means now to be "amid the Anthropocene." One might point out that all of this theology has been forged "in" the time of the Anthropocene, even with the late start date assigned by the Anthropocene Working Group—namely, the early 1950s. But part of the salience of thinking theologically "amid the Anthropocene" seems to be about *becoming conscious* of this positioning; if such consciousness is hinged to the nominalization of a new epoch, then this of course has come later, since the early 2000s. (That there is such an epistemological gap should itself be an important reference point in any reckoning of the significance of the "Anthropocene.") So, beyond mere temporal overlap, I would argue that in regard to some of the most substantive questions thrown up by a consciousness of being in this new geohistorical epoch, there are important theological anchors present, not only in Schillebeeckx's work but also in political, liberation, and ecological theologies more broadly.

In order to identify these points of contact, I will return to the "Anthropocene" story presented above to explore how its account of qualitative change splinters and stalls. We saw that consideration of the anthropogenic causes of Earth System change is tied to an understanding of a radically altered relationship between humanity and nature, which in turn is tied to a sense that the character and extent of creaturely suffering have been transformed. These overlapping claims are key to the assertion that the circumstances in which Christians now seek to make sense of God's providence are unprecedented. The upshot of revisiting these claims as places of narrative difficulty is not to deny the story outright but rather to suggest that an "Anthropocene" consciousness will benefit from tempering its singularity with respect to lessons learned about discerning God's loving care amid overwhelming distress.

■ Cracks and Questions

At first blush, the crux of the maximalist's narrative arc is an inflated display of human agency, which leads to the suffering of nonhuman nature and ultimately our own as well. The error in this behavior hinges on a mistreatment of "nature," which is to say, more precisely, a pattern of interventions that profoundly disrupt the stability of the Earth System. This technologically-charged, fossil-fueled human activity bespeaks a "power-over" that might be taken to challenge traditional notions of God's providence in its capacity to change history and nature alike. Indeed, for some ecomodernists this is just the realization needed: how much is in our grasp, and our high calling to embrace the charge of a "good Anthropocene."[67] What quickly shows through from behind the veneer of novelty, though, is a more familiar refrain of human exceptionalism. To my mind, this observation holds whether the proponents in question advocate a very early "Anthropocene" start date, which makes it basically coterminous with human existence, or hew to a teleological chronology of progress in which we have just lately ascended properly into our role as "gods"—and a good thing![68] Thus for a scholar like Eileen Crist, the problem with "the Anthropocene" is its ongoing anthropocentrism; *this* version of human responsibility, she contends, replays rather than erodes the fantasy of humanity's separation from and control of nature.[69] For her and others, the result might be an asterisk concerning how new, really, is this picture of the disastrous agency of "the Anthropos," given the extent to which it relies on and buoys an older story of human power.

It could be countered that talk of human power is simply descriptive. Yet not only moral evaluations of human activity in the "Anthropocene" but also interpretations of the very "shift" at issue spill out in opposite directions. While some commentators, as we have seen, speak of a "human-dominated planet" (Lewis and Maslin), others are more hesitant. Noel Castree, for instance, seeks to clarify that speaking of an anthropogenic epoch "is not to imply that humans somehow 'dominate' or 'control' the earth, but simply to acknowledge their newfound capacity to alter biophysical 'boundary conditions' across multiple large-scale environmental systems."[70] More striking still, others observe that if the "Anthropocene" names humankind as *cause* then in the next breath one must acknowledge it as *subject*.[71] The agent of immense suffering is the patient of it, too. Any talk of power must be paired with the evidence (piling up daily, it seems) of a loss of control. The consequences at hand are not just troubling but spiraling, cascading, and nonlinear. To what extent, then, does this new epoch hurl us back to the age-old problem of humans' struggle for survival within—or even against?—a nature that cannot be tamed?[72] Eva Horn and Hannes Bergthaller identify the "specific structure of human agency in the Anthropocene, which consists of a paradoxical combination of

---

67. Asafu-Adjaye et al., "An Ecomodernist Manifesto."

68. See Lynas, *The God Species*.

69. See Crist, "On the Poverty of Our Nomenclature."

70. Castree, "The Anthropocene and Geography."

71. Horn and Bergthaller, *The Anthropocene*, 67.

72. See, e.g., Hamilton, *Defiant Earth*; Horkheimer and Adorno, *Dialectic of Enlightenment*.

enormous force and an equally profound loss of control" as one of its defining features.[73]

For theists, Horn and Bergthaller's conclusion should suggest that a story in which human agency has self-evidently overshadowed divine action in the world is premature. Then again, we didn't need to be amid the "Anthropocene" to know that such thinking falls short of Christianity's richer insights regarding "the interplay between divine, human and other forms of agency"—that is, *concursus*—which, as the editors' introduction aptly notes, "suggest[s] convivence and accompanying, but not control, determinism, or domination." This is not to say that what the "Anthropocene" throws up does not test Christian notions of providence, but only that such challenges take root in an already rigorous field of theological reflection on this topic. Philosophical and systematic theologians have articulated notions of "double agency"[74] and, like Schillebeeckx, a "non-competitive" or "non-contrastive" relationship between God and creation.[75] In the dialogue between theology and evolutionary science and cosmology, the depth of reflection on secondary causality as the mode of God's providential work in the world has been especially productive as the broader sweep of "all creation" comes into view.[76] Simplistic "God of the gaps" thinking should not gain traction given the robust alternatives on offer. Nonetheless, present planetary conditions may well instigate renewed negotiations of the layering of multiple forms of causal power—especially when the "paradoxical" structure of human agency appears front and center. Such negotiations lead to questions not only about restraining obviously destructive human behavior/systems but also about interpreting humanity's lack of control over its environment. Is the latter an unadulterated threat? And, if so, does this too easily revive a narrative of God joining humankind's battle against the natural world? Would it kindle multispecies solidarity or make it more difficult to hold that "God's charity is broad enough for bears"?[77] Is it an opportunity for a more humble self-assessment, or even a stark confrontation with the ruin wrought by the hubristic regime of endless growth? In that case, might God's activity be most aligned with the chaotic volatility of the "Anthropocene"? Is there divine justice to be found in the planet's biting back? Ernst Conradie's essay on divine *conservatio* suggests

---

73. Horn and Bergthaller, *The Anthropocene*, 72.

74. See Farrer, *Faith and Speculation*, with thanks to Ernst Conradie for the reference.

75. See Tanner, *God and Creation*.

76. See Edwards, *How God Acts*, especially his discussion of the problem of evolutionary suffering (chapter 1) and his "noninterventionist theology," which insists (much like Farrer) that we do not know *how* God acts through secondary causes (chapter 4). See also Johnson, "Does God Play Dice?" and Haight, "Spirituality, Evolution, Creator God."

77. John Muir, as quoted by Johnson, *Creation and the Cross*, 28.

that this terrible prospect is one that must at least be raised—if only to be put back on the shelf—in the search to understand where God's care is evident and efficacious. On balance, the contours of this search today are distinct but not qualitatively different from those pursued in ecotheology over the past several decades.

Closely related to the issue of agency, one can see from another angle how the language that describes the putatively novel status of human beings contains the seeds of its own interrogation. Earlier, we looked at several representative examples of how best to describe the shifting positionality of "human" relative to "nature." Consider as well Noel Castree's summary claim that the "Anthropocene" "describes not merely the 'human impact' on the nonhuman world but also the folding of human activity into Earth-surface systems such that it becomes in some sense endogenous to those systems."[78] Even as claims about what differentiates the "Anthropocene" appear on one level to accentuate the separation of humanity from nature through humanity's causal power-over, they reveal a proximity, an involvement, an entanglement—to use a term popular in posthumanist planetary thinking. The relational index that is meant to register the uniqueness of this new epoch shows itself to be unstable: both difference from and likeness to "nature" are required to make sense of the human status in the "Anthropocene," but each characterization tugs at the heels of the other and threatens its coherence.

For some environmental theorists, the reaction to this apparent dilemma is the equivalent of a knowing "I told you so." Bruno Latour is a prime example. For Latour, there is nothing "good" about the "Anthropocene" in the way that the ecomodernists mean it.[79] There is, however, something akin to the revelation of a buried truth through the ontological reconfiguration it effects over and against a foundational ideology of Western modernity: namely, the divide of (human) culture, politics, and history from (nonhuman) nature.[80] If the "Anthropocene" is taken to be coterminous with or issuing from "modernity," then its characterization as a radical rupture reveals the same illusory quality: we moderns have never been modern and we humans have never been anything but part of nature.[81] And yet not nature in the hollowed-out form imposed on it by modern Western science: Latour is not interested in capitulating to this dualism by collapsing it into one of its terms (nature vs. culture) but rather overcoming

---

78. Castree, "The Anthropocene and Geography."

79. See Latour's response to ecomodernism, "Fifty Shades of Green."

80. Latour, "Anthropology at the Time of the Anthropocene."

81. Latour, *We Have Never Been Modern*.

the binary by disintegrating and reconfiguring these terms (viz., into "distributed agencies" and a "down to earth" politics).[82]

Latour's work is distinctive, but it is also representative of an overarching concern in environmental thought, including ecotheology. Negotiations of human belonging (in/as nature) as a crucial hinge of responsibility (to/for nature) have arguably been the central thread in ecotheology and in environmental thought more broadly for the past five to six decades. Through various combinations of biblical, doctrinal, philosophical, ethical, and scientific inquiries, scholars have engaged in a quest for viable alternatives to the portrait of "Man" around which a culture-nature dualism took shape and brought with it so much ecological devastation. This has meant a process of sifting—with tools of critique, recovery, and reformulation—among Christian and cultural sources tied, whether rightly or wrongly, to dualistic thinking: from the Bible to medieval scholasticism and the Protestant Reformation; from Cartesian and Enlightenment philosophy to the scientific and industrial revolutions. Projects of extension and expansion comprise no small share of ecotheology's reconnective work at the site of this divide: on the one hand, all that exists is natural or dependent on nature; on the other, it is not humanity alone that possesses traits such as subjectivity, agency, dignity, and rights.

So too, and more challenging yet, other scholars deconstruct these cherished ideals and even take aim at the concept of "humanity" itself, arguing not merely for its circumstantial withholding from the racialized—and often gendered—subjects created by colonialism and the trans-Atlantic slave trade, but for that suppression as constitutive of the concept's value for Europeans and their white inheritors. Such work has not always been seen as relevant to ecotheological debates, but latterly a number of scholars, especially scholars of color, have made its connections to the domain of human-nature relations increasingly clear. They point to the liabilities of rehabilitating this burdened notion of a universal human "we" in the context of environmental and climate crises wherein "humanity" is said to be both in need of healing (to correct its pattern of exploiting the natural world) and profoundly under threat (by a warming and unpredictable climate). Such critical perspectives are all the more trenchant in light of the deeply uneven contributions to and consequences of regional and global environmental-climatic problems.[83] Meanwhile, in recent years scholars of race and coloniality have elucidated links between the experiences of historically dominated peoples in the Americas and nondominating ways

---

82. See Latour, *Down to Earth*.

83. There is a substantial, although less than complete, overlap between these largely North American theoretical paradigms and calls for ecojustice arising from the Global South.

of relating to the Earth.[84] These efforts are far from a romantic retreat into Nature (an enduring temptation in ecotheology). Rather, they also trouble that pole of the binary as the twinned "other" of normative humanity. Their strategies of analytical unpicking and creative reimagination cover a wide range of perspectives, some of which resonate with posthumanist and new materialist insights while others chart a different course through the ruins of anthropocentric eco-social disaster.

With respect to this already complex and fecund discussion around "human" and "nature" in ecotheology, the "Anthropocene" signals neither the introduction nor the resolution of key questions about these terms' meaning and relationship but lies squarely in their midst. While consciousness of the transition into a new global epoch has instigated a period of academic generativity and cultural-spiritual soul-searching in regard to its titular subject, the *Anthropos*, I would argue that it replays and reshuffles well-established elements in environmentalist thought. Even as theologians and theorists venture new vocabularies of being, and of being human, some reserve is fitting: their work also indicates the enduring hold of some dichotomized sense of alterity and connection in regard to the other-than-human world. The explosion of "Anthropocene" discourse surrounding its basic premise of novelty—human impact *on* the Earth such that humans have become a force *of* nature—only drills down further into this territory. As with the paradox of agency, Horn and Bergthaller (among others) take this to be an insuperable point of tension: describing the "Anthropocene" "implies seeing the human as a being that is at once natural and unnatural, both a part of nature and apart from nature."[85] Their conclusion echoes that of many authors in ecotheology who find that for all their wrestling with the nature–culture divide—contingent and illusory as it may be—they must settle for a grammar of human–nature relations along these "both/and" lines. Unsatisfying, perhaps, but necessary.

■ From the Aporia: Ethics and Providence

One reason for dwelling on these difficulties within claims about the "Anthropocene" as an utter rupture is that—as the section above shows—the task of description is charged with prescriptive import. This, of course, is a foundational insight of twentieth-century critical thought. As I suggested at the outset, it is one that has become an essential tool for theologians reflecting on the reality of suffering and the function of the doctrine of providence. They have cultivated awareness of the way that

---

84. See, among many: Jennings, "Reframing the World"; Harris, *Ecowomanism*; Whyte, "Indigenous Climate Change Studies."

85. Horn and Bergthaller, *The Anthropocene*, 67.

judgments of divine action—present here, absent there—carry moral weight and participate in the perpetuation or, possibly, interruption of patterns of worldly violence. What I wish to emphasize in addition is that the ethical analysis must also flow the other way. Narratives about what God is doing in the world that seek to understand the relative suffering and responsibility (agency) of various entities—i.e., humanity and nature—will be shaped to some degree by those concepts' normative associations. The previous section traced some of these associations as they have been rekindled by discussion of the "Anthropocene." These concepts may even imply some notion of providence in the boiled-down sense of conveying what ought to be or what must be.

The idea that Christian notions of providence traverse secular contexts and discourses is a well-trod path.[86] Beyond those places where God's governance of history is explicitly defended, commentators see its basic form reflected in narratives of history or human rational ability as the ascendant march of progress. Clive Hamilton indicts the *Ecomodernist Manifesto* for one such optimistic view, which carries with it the "essential political and moral flaw" of a stultifying and pacifying perspective on the suffering wrought by ecological damage.[87] With anthropological leanings that diverge from Hamilton's call for a "new anthropocentrism," Latour launches a similar criticism regarding the supposedly providential arc of the "good Anthropocene."[88] His charge that the document's authors refuse to recognize human beings' real entanglement with nature obtains more broadly in a range of eco-managerial approaches, from more conservative versions of stewardship[89] to technocratic and transhumanist approaches.[90]

That what is covered over (by way of legitimating justification) includes both the suffering of nonhuman nature and human suffering—especially of the poor and already vulnerable—reflects an important fact. As the previous section labored to show, an idea of humanity as natural/unnatural cannot be entirely dispensed with, and yet it is not supportable in any firm way. Ecotheology roundly demonstrates this point; "Anthropocene" discourse ups the ante. From the other side, then, the traditional bifurcation of the *causes* of suffering into natural and social categories should receive similar

---

86. Both Ernst Conradie's and Clive Pearson's essays in this volume engage secularized versions of providence.

87. Hamilton, "The Theodicy of the 'Good Anthropocene'."

88. Latour writes to ecomodernists: "And please don't tell me that you have no enemies, and that it is all about tracing the obvious and inevitable path of reason and progress—because I know who has drawn that path. It is a providential God, which is not my God" ("Fifty Shades of Green").

89. See Horan, *All God's Creatures*.

90. See Neyrat, *The Unconstructable Earth*.

scrutiny—as indeed it has.⁹¹ What this means is that there is also a need for alertness to the way in which problematic renderings of providence can inhere where "nature" (rather than Man or history) has the lead role in the story or equally where some combinatory or unitary vision is put forward.

A wide range of humanities scholars have approached the notion of humanity as a species-agent and as a geological force in this light, albeit without using the language of providence. They have done so as a corrective to at least two implications of the early scientific expositions of this new geohistorical epoch and of the *Anthropos* at its center. These narrative consequences may not have been intentional, and there is a case to be made that scientific expositors have sought to take critical perspectives on board,⁹² but their stakes are high. In short, the object of scrutiny is a naturalizing account of the radical shift in question, which is important to examine for its bearing on the status of anthropogenic causes and the issue of moral responsibility.

The first criticism is that the generalized species-concept "obscures the concrete social and economic arrangements that have destabilized the Earth System."⁹³ It is especially relevant when ecological degradation, climate change, and the arrival of the "Anthropocene" are figured as *unintentional* effects of human activity—of human nature.⁹⁴ Even if "we" as a human species are also subject to nature, if we are well and truly part of nature (as an evolved biological species), is the suffering we inflict rendered inevitable, or in any case thereby excused or rationalized as being outside the realm of moral evaluation? What good is it to argue with a "force of Nature" or to try to change its course? This goes against the grain of the environmentalist concern that seems to have formed part of the push toward designating the "Anthropocene," yet seems also to be a kind of gravitational pull within the narrative structure. Moreover, critics have observed how the naturalizing bent can ricochet this species-concept of humanity back into the realm of an anthropocentric and technocratic project. Drawing on Lisa Sideris's work, Willis Jenkins writes, "Climate engineering and the Anthropocene became reflexively warranting, the idea in one driving a sense of inevitability in the other [... T]he Anthropocene

---

91. See the editors' introduction for a helpful outline of traditional views.

92. So argues Clive Pearson in his essay in this volume.

93. Horn and Bergthaller, *The Anthropocene*, 69

94. Malm and Hornberg observe: "The 'Anthropocene' registers this moment of epiphany: the power to shape planetary climate has passed from nature into the realm of humans. As soon as this is recognised, however, the main paradox of the narrative, if not of the concept as such, becomes visible: climate change is *denaturalised* in one moment—relocated from the sphere of natural causes to that of human activities—only to be *renaturalised* in the next, when derived from an innate human trait, such as the ability to control fire. Not nature, but human nature—this is the Anthropocene displacement" ("The Geology of Mankind?" 65).

idea can make collective terraforming into a scientific fact, confirmed by geological periodization, in a way that extends the enterprise of power while exempting it from critique."[95] For those who see the "Anthropocene" as a validation of interconnectedness—maybe even as a decisive overcoming of the split between nature and history—such observations should give pause.

What I wish to stress once again is how an "Anthropocene" consciousness accentuates but does not originate these weighty questions: How is "human" to be related to "nature"? And what ethical options are enabled or constrained by one configuration or another? Because of the obvious cost of the "great divide" between nature and culture—a fundamentally anthropocentric ideology—much of the work of ecotheology and science-engaged theology has been to cultivate models of belonging and creaturely community. The sometimes stated, often assumed premise is that a felt sense of belonging will spur respect, care, and responsibility for and with our creaturely kin and our earthly or cosmic home.[96] Without necessarily rejecting that premise wholesale, other ecotheologians have argued for chastened forms of human distinctiveness—often to counter perceived threats to the struggle for justice for marginalized people, but also in some cases for the good of the Earth System itself.[97] The debate remains live. As sketched in the previous section, ongoing responses to Enlightenment-industrial-colonial "humanity" continue to be contested in the field that bears this name. Used as a blunt instrument, denaturalizing critique can re-entrench an unhelpful binary with tit-for-tat allegations (any turn toward the "natural" is bad). A finely-tuned version, however, is essential for disclosing ideological distortions that crowd out alternative possibilities for living together, within and across species boundaries. Insofar as these distortions are the source of so much unnecessary suffering, "denaturalizing Anthropocene myths"—as Jenkins puts it—must be one step in making room to tell the story of how God may be acting here, "amid the Anthropocene."

This concern about a naturalizing narrative frames a second line of criticism directed at the Anthropos: its homogenizing function. The contingent social and economic systems identified behind the veil of "anthropogenic" causes (see: colonialism, capitalist extractivism, and the fossil-fuel economy) are arrangements designed to serve particular

---

95. Jenkins, "Ethics after Humanity," 616.

96. Examples are manifold. Pope Francis provides one: "Because all creatures are connected, each must be cherished with love and respect" (*Laudato Si'*, 42).

97. See Castillo, *An Ecological Theology of Liberation* and Sideris, *Environmental Ethics, Ecological Theology, and Natural Selection*.

interests and which continue, in large part, to do so. To ignore the striking inequalities and asymmetries operative in both the causes and consequences of Earth System change is to indulge a fantasy that generic humanity anywhere exists. That is not a neutral error but a grave injustice. Heather Davis and Zoe Todd make one powerful intervention along these lines, arguing that "the Anthropocene continues a logic of the universal which is structured to sever the relations between mind, body, and land"; this is a logic that "in its reassertion of universality, implicitly aligns itself with the colonial era," an era whose appropriative and exploitative dynamics they spell out.[98] Jenkins powerfully summarizes these voices and others in his insistence that religious ethics in the time of the "Anthropocene" must, in effect, denaturalize its undifferentiated *anthropos*:

> The humanity which must steward planetary systems actually names particular social forms that have already appropriated much of the biosphere for themselves, first by rendering its relations into "resources" and then by deploying extractive colonialism and racial capitalism to take them. Meanwhile the planetary consequences of that appropriation are anonymized as the fault of humanity in general, which erases the historical violence from which the climate crisis emerges and launders the resulting intensification of inequalities into a predicament for the species. Humanity can function as a laundering concept, assigning both causes and consequences of climate change to the species in general, rendering invisible the historical and structural violence from which it is produced and experienced.[99]

Here again, to denaturalize means to recognize the historicity and partiality of humanity as a category and to confront the violence (which is always also to say suffering) that it "at once authorize[s] and conceal[s]," in both the past and the present.[100]

This effects an important shift, or in any case a clarification, with respect to the claim that the "Anthropocene" (or consciousness thereof) has utterly transformed the human condition and therefore also the problem of creaturely suffering. Guided by Schillebeeckx's anti-theodical stance and buoyed by more recent articulations of the dangers of naturalizing and homogenizing "humanity" as a category, one might bring their insights to the volume's guiding question in this way: suffering is already maximally difficult. While there is, no doubt, something salient and shocking about the scale at issue in the "Anthropocene's" geological and system-level

---

98. Davis and Todd, "On the Importance of a Date," 761.

99. Jenkins, "Ethics after Humanity," 619.

100. Jenkins, "Ethics after Humanity," 619. I would note again that denaturalizing does *not* mean resigning oneself to the terms of the "the great divide." To the contrary, when rightly used it helps to see how that modern construction is shaped by coloniality and racial capitalism, and therefore how genuine alternatives—including perhaps a notion of "planetary humanity"—can emerge from neglected archives. Jenkins's engagement with theorist of the Black Atlantic Paul Gilroy is instructive here (see 621–22).

ruptures, adopting that vantage point for commentary on human meaning and divine purpose runs a great risk. The refusal of the privileged, or simply those of us in the present, to accord the suffering-unto-death of certain groups the finality it merits—to see it as sufficiently cataclysmic—is a grave moral and spiritual failing. For Schillebeeckx, "critical remembrance of suffering humanity" is essential to unleashing the critical and productive force of creation faith for a better (a possible) future.[101] And in that respect, talk of providence is not a one-size-fits-all story.

In advancing these lines of critique, I want to insist that one is not thereby committed to saying "nothing has changed," or in other words to siding with the humanities over and against the maximalist interpretation characteristic of the natural sciences, and thus failing to appreciate the existential significance of a rupture in the Earth System. The disciplinary debate has arisen from substantive differences in perspective and priorities. "However," as Horn and Bergthaller contend, "this polemic opposition between an undifferentiated 'species thinking' and a properly discriminating historical perspective has in fact confused rather than clarified the issue."[102]

For them, the much-discussed contribution of Dipesh Chakrabarty is helpful in more clearly navigating the complexity at hand. They note that while he has "occasionally been consigned to the 'naturalist' camp," he has in fact "argued that the real challenge of thinking about the human in the Anthropocene consists precisely in the 'collision' of different conceptions of the human."[103] Where I have employed the generic and universalizing language of the standard "Anthropocene" narrative in this essay, it is in part to show how, even on its own terms, it fractures along a series of fault lines which point us back to rather than away from key issues in humanities. As Jenkins maintains, theology in particular "may become increasingly significant for the cultural work of interpreting humanity's new roles with the planet."[104] More pointedly, one might hold that references to the facticity of irreversible ecological, planetary, and/or Earth System change should inspire more rather than fewer of the critical questions around power, normativity, and injustice honed by contemporary theology and other humanities disciplines. Per Jenkins: "One task for ethics thus remains the same as ever: to track interactions of moral and empirical cosmology and open them to reflective criticism."[105] Even so, he also affirms

---

101. See especially Schillebeeckx, *Christ*, 663–714.

102. Horn and Bergthaller, *The Anthropocene*, 69.

103. Horn and Bergthaller, *The Anthropocene*, 69.

104. Jenkins, "Ethics after Humanity," 618.

105. Jenkins, "Ethics after Humanity," 617. He writes: "Some Anthropocene pronouncements permit climate crisis to overdetermine moral conditions. Humanity has always been in peril. As [Charles] Long's works

without hesitation Chakrabarty's fundamental demand to see "that the Anthropocene idea destabilizes premises by which research in the humanities has been organized."[106] This is crucial: neither disciplinary perspective nullifies the truth of the other.[107]

■ A Coda on Creation Faith

Where does this leave us? A clarity borne of "collision" is not a resolution. But it may permit a more assiduous perseverance to remain in the genuine aporia that is the "Anthropocene." Inspired by Schillebeeckx's rendering of creation faith, I am inclined to see such perseverance as a performative enactment of a "basic trust" that may be at once existential and religious: an experience of the absolute presence of the creating and saving God. And, as I have suggested in this essay, as a way of making room to speak about this God at the confluence of new and not-new paths of creaturely suffering. There is much in Schillebeeckx's thought that bears further consideration in light of the main predicaments thrown up by the "Anthropocene," including his understanding of human finitude in relation to contemporary questions of human agency and his burgeoning interest, late in his career, in humanity's co-creatureliness.[108] In response to the present volume's focus on suffering, the places where content bleeds into method are especially instructive. Critical negativity names one of these places. Critique on behalf of those who suffer represents an essential element of what it means to demonstrate trust in a loving God who desires the good and flourishing of the poor, the downtrodden, and of all creation. It is not merely the prolegomenon to the Christian story, Schillebeeckx would say, but rather a central thread. Here amid the "Anthropocene," we cannot do without it.

(footnote 105 continues)
probe, modernity's humanity was in crisis from its formation in service to the Atlantic slave trade and the simultaneous creation of 'another basis for the human […] from the bowels of a slave ship'. […] Can religious ethics incorporate terraforming agencies and planetary perils into its analyses without the assumption that this moment epitomizes the human species?" (611–12).

106. Jenkins, "Ethics after Humanity," 613.

107. See a parallel claim from Horn and Bergthaller: "The core argument of this chapter [titled 'The *Anthropos*'] is that these two perspectives on the human in the Anthropocene should be seen as addressing two complementary conceptions of the human: the first as a cultural and social being, the second as a biological species. In order to understand the role and the ambiguous structure of human agency in the Anthropocene as intentional power and unintentional force (Chakrabarty 2018), we believe, one needs to view these two conceptions as standing in an irresolvable but indispensable tension" (*The Anthropocene*, 68). However, their description of the former as "a politically, culturally, and economically differentiated image of humanity" is subsequently overshadowed when this humanities-led conception is then associated with ecomodernism, which the authors say "absolutizes the idea of the human as *homo*—a rational, self-determining being capable of making and carrying out ethical decisions" (69, 70). Yet it would be a strain to characterize ecomodernism as a "properly discriminating historical perspective."

108. See Pyne, "Reading Schillebeeckx in an Ecological Age"; Poulsom, "Concentrating on Creation."

■ Bibliography

Asafu-Adjaye, John, et al. "An Ecomodernist Manifesto." http://www.ecomodernism.org/manifesto-english/.

Boff, Leonardo. *Cry of the Earth, Cry of the Poor*. Translated by Phillip Berryman. Maryknoll: Orbis, 1997.

Bonneuil, Christophe, and Jean-Baptiste Fressoz. *The Shock of the Anthropocene: The Earth, History, and Us*. Translated by David Fernbach. London: Verso, 2017.

Castillo, Daniel P. *An Ecological Theology of Liberation: Salvation and Political Ecology*. Maryknoll: Orbis, 2019.

Castree, Noel. "The Anthropocene and Geography." In *Oxford Bibliographies*, edited by Brian Wharf, 1–20. Oxford: Oxford University Press, 2017. https://doi.org/10.1093/obo/9780199874002-0111

Claesson, Bo. "Edward Schillebeeckx (1914-2009)—In Search of a Better Tomorrow." In *Creation and Salvation Volume 2: A Companion on Recent Theological Movements*, edited by Ernst M. Conradie, 82–85. Zürich: LIT Verlag, 2012.

Cohen, Tom, J. Hillis Miller, and Claire Colebrook. *Twilight of the Anthropocene Idols*. London: Open Humanities Press, 2016.

Conradie, Ernst M. *Secular Discourse on Sin in the Anthropocene: What's Wrong with the World?* Lanham: Lexington Books, 2020.

———. "Some Theological Reflections on Multi-Disciplinary Discourse on the 'Anthropocene'." *Scriptura* 121 (2022) 1–23. https://doi.org/10.7833/121-1-2076

Cooper, Jennifer. *Humanity in the Mystery of God: The Theological Anthropology of Edward Schillebeeckx*. London: T & T Clark, 2009. Reprinted 2011.

Crist, Eileen. "On the Poverty of our Nomenclature." *Environmental Humanities* 3:1 (2013) 129–47. https://doi.org/10.1215/22011919-3611266

Crutzen, Paul, and Will Steffen. "How Long Have We Been in the Anthropocene Era?" *Climatic Change* 61 (2003) 251–57. https://doi.org/10.1023/B:CLIM.0000004708.74871.62

Davis, Heather, and Zoe Todd. "On the Importance of a Date, or Decolonizing the Anthropocene." *Acme: An International Journal for Critical Geographies* 16:4 (2017) 761–80.

Descola, Philippe. *Beyond Nature and Culture*. Translated by Janet Lloyd. Chicago: University of Chicago Press, 2013.

Edwards, Denis. *How God Acts: Creation, Redemption, and Special Divine Action*. Minneapolis: Fortress, 2010.

Farrer, Austin. *Faith and Speculation: An Essay in Philosophical Theology*. London: A & C Black, 1967.

Francis (Pope). *Laudato Si'. Encyclical Letter on Caring for our Common Home*. Vatican: Vatican Press, 2015. https://www.vatican.va/content/francesco/en/encyclicals/documents/papa-francesco_20150524_enciclica-laudato-si.html.

Haight, Roger. "Christ and Culture." In *T & T Clark Handbook of Edward Schillebeeckx*, edited by Stephan van Erp and Daniel Minch, 395–409. London: T&T Clark, 2019.

———. "Spirituality, Evolution, Creator God." *Theological Studies* 79:2 (2018) 251–73. https://doi.org/10.1177/0040563918766717

Hamilton, Clive. *Defiant Earth: The Fate of Humans in the Anthropocene*. Cambridge: Polity, 2017.

———. *Requiem for a Species: Why We Resist the Truth about Climate Change*. London: Earthscan, 2010.

———. "The Theodicy of the 'Good Anthropocene'." *Environmental Humanities* 7:1 (2016) 233–38. https://doi.org/10.1215/22011919-3616434

Haraway, Donna J. *Staying with the Trouble: Making Kin in the Chthulucene*. Durham: Duke University Press, 2016.

Harris, Melanie L. *Ecowomanism: African American Women and Earth-Honoring Faiths*. Maryknoll: Orbis, 2017.

Hilkert, Mary Catherine. "Experience and Revelation." In *The Praxis of the Reign of God*, 2nd ed., edited by Mary Catherine Hilkert and Robert J. Schreiter, 59–78. New York: Fordham University Press, 2002.

———. "The Story of Jesus and Human Flourishing: Schillebeeckx's Later Soteriological Anthropology." In *T & T Clark Handbook of Edward Schillebeeckx*, edited by Stephan van Erp and Daniel Minch, 278–98. London: T&T Clark, 2019.

Hilkert, Mary Catherine and Robert Schreiter, eds. *The Praxis of the Reign of God: An Introduction to the Theology of Edward Schillebeeckx*. 2nd ed. New York: Fordham University Press, 2002.

Hill, William J. "A Theology in Transition." In *The Praxis of the Reign of God*, 2nd ed., edited by Mary Catherine Hilkert and Robert J. Schreiter, 1–18. New York: Fordham University Press, 2002.

Horan, Daniel. *All God's Creatures: A Theology of Creation*. Lanham: Lexington Books/Fortress Academic, 2018.

Horkheimer, Max, and Theodor W. Adorno. *Dialectic of Enlightenment*, edited by Gunzelin Schmid Noerr. Translated by Edmund Jephcott. Stanford: Stanford University Press, 2007.

Horn, Eva and Hannes Bergthaller. *The Anthropocene: Key Issues for the Humanities*. London: Routledge, 2019.

IPCC. "Summary for Policymakers." In *Climate Change 2021: The Physical Science Basis. Contribution of Working Group I to the Sixth Assessment Report of the Intergovernmental Panel on Climate Change*, edited by V.P. Masson-Delmotte et al., 3–32. Cambridge: Cambridge University Press, 2021. https://doi.org/10.1017/9781009157896.001

Jenkins, Willis. "Ethics After Humanity." *Journal of Religious Ethics* 51:4 (2023) 611–38. https://doi.org/10.1111/jore.12457

Jennings, Willie James. "Reframing the World: Toward an Actual Christian Doctrine of Creation." *International Journal of Systematic Theology* 21:4 (2019) 388–407. https://doi.org/10.1111/ijst.12385

Johnson, Elizabeth A. *Creation and the Cross: The Mercy of God for a Planet in Peril*. Maryknoll: Orbis, 2018.

———. "Does God Play Dice? Divine Providence and Chance." *Theological Studies* 57:1 (1996) 3–18. https://doi.org/10.1177/004056399605700101

———. *Quest for the Living God: Mapping Frontiers in the Theology of God*. New York: Bloomsbury, 2011.

Kennedy, Philip. *Schillebeeckx*. Collegeville, MN: Liturgical Press, 1993.

Kirsch, Adam. *The Revolt Against Humanity: Imagining a Future Without Us*. New York: Columbia Global Reports, 2023.

Kolbert, Elizabeth. *The Sixth Extinction: An Unnatural History*. New York: Henry Holt, 2014.

Latour, Bruno. *Anthropology at the Time of the Anthropocene: A Personal View of What Is to Be Studied*. Washington, DC: Distinguished Lecture to American Association of Anthropologists, 2014. http://www.bruno-latour.fr/sites/default/files/139-AAA-Washington.pdf.

———. *Down to Earth: Politics in the New Climatic Regime*. Translated by Catherine Porter. Cambridge: Polity, 2018.

———. "Fifty Shades of Green." July 3, 2015. http://modesofexistence.org/on-breakthrough-dialogue-2015-the-good-anthropocene-june-21---23-2015/.

———. *We Have Never Been Modern*. Translated by Catherine Porter. Cambridge: Harvard University Press, 1993.

Lee, Michael E. "Schillebeeckx and the Path to a Liberation Theology." In *T & T Clark Handbook of Edward Schillebeeckx*, edited by Stephan van Erp and Daniel Minch, 410-23. London: T&T Clark, 2019.

Lewis, Rhona. *Concentrated Creation: Creation and Salvation in the Christology of Edward Schillebeeckx*. London: T&T Clark, 2023.

Lewis, Simon, and Mark Maslin. *The Human Planet: How We Created the Anthropocene*. New Haven: Yale University Press, 2018.

Lynas, Mark. *The God Species: Saving the Planet in the Age of Humans*. Washington, DC: National Geographic, 2011.

Malm, Andreas, and Alf Hornberg. "The Geology of Mankind? A Critique of the Anthropocene Narrative." *The Anthropocene Review* 1:1 (2014) 62-69. https://doi.org/10.1177/2053019613516291

McManus, Kathleen Anne. *Unbroken Communion: The Place and Meaning of Suffering in the Theology of Edward Schillebeeckx*. Lanham: Rowman & Littlefield, 2003.

Neyrat, Frédéric. *The Unconstructable Earth: An Ecology of Separation*. Translated by Drew S. Burk. New York: Fordham University Press, 2019.

Nixon, Rob. *Slow Violence and the Environmentalism of the Poor*. Cambridge: Harvard University Press, 2018.

O'Meara, Janet M. "Salvation: Living Communion with God." In *The Praxis of the Reign of God*, 2nd ed., edited by Mary Catherine Hilkert and Robert J. Schreiter, 97-116. New York: Fordham University Press, 2002.

Poulsom, Martin. "Concentrating on Creation: Following Christ in a Context of Climate Change." In *Grace, Governance and Globalization*, edited by Stephan van Erp, Martin G. Poulsom, and Lieven Boeve, 125-39. T & T Clark Studies in Edward Schillebeeckx. London: Bloomsbury, 2017.

———. "Schillebeeckx's Praxis of Creation." In *T & T Clark Handbook of Edward Schillebeeckx*, edited by Stephan van Erp and Daniel Minch, 237-51. London: T&T Clark, 2019.

Pyne, Elizabeth M. "Reading Schillebeeckx in an Ecological Age: Creation Faith and the Politics of Nature." In *T & T Clark Handbook of Edward Schillebeeckx*, edited by Stephan van Erp and Daniel Minch, 436-56. London: T&T Clark, 2019.

Rego, Aloysius. *Suffering and Salvation: The Salvific Meaning of Suffering in the Later Theology of Edward Schillebeeckx*. Louvain: Peeters, 2006.

Riley, Matthew T., and Whitney A. Bauman. "Wicked Problems in a Warming World: Religion and Environmental Ethics." *Worldviews* 21:1 (2017) 1-5. https://www.jstor.org/stable/26552270.

Rockström, Johan, et al. "Planetary Boundaries: Exploring the Safe Operating Space for Humanity." *Ecology and Society* 14:2 (2009). http://www.ecologyandsociety.org/vol14/iss2/art32/.

Schillebeeckx, Edward. *Christ: The Christian Experience in the Modern World*. London: T&T Clark, 2014. Reprinted 2018. [Dutch original: 1977].

———. *Church: The Human Story of God*. London: T&T Clark, 2014. Reprinted 2018. [Dutch original: 1989]

———. *For the Sake of the Gospel*. Translated by John Bowden. London: SCM, 1989.

———. *God Among Us: The Gospel Proclaimed*. Translated by John Bowden. New York: Crossroad, 1983.

———. *I Am a Happy Theologian*. Translated by John Bowden. New York: Crossroad, 1994.

———. *Interim Report on the Books Jesus and Christ*. London: T&T Clark, 2014. Reprinted 2018. [Dutch original: 1980].

———. *The Language of Faith: Essays on Jesus, Theology, and the Church*. Maryknoll: Orbis, 1995.

———. *The Understanding of Faith: Interpretation and Criticism*. Translated by N. D. Smith. New York: Seabury, 1974.

Scott, Peter. "The Re-Homing of the Human: A Theological Enquiry into Whether Humans Are at Home on Earth." In *Christian Faith and the Earth: Current Paths and Emerging Horizons in Ecotheology*, edited by Ernst M. Conradie, Sigurd Bergmann, Celia Deane-Drummond, and Denis Edwards, 115–36. London: T&T Clark, 2014.

Sideris, Lisa H. *Environmental Ethics, Ecological Theology, and Natural Selection*. New York: Columbia University Press, 2003.

Steffen, Will, et al., "The Anthropocene: Conceptual and Historical Perspectives." *Philosophical Transactions of the Royal Society A* 369:1938 (2011) 842–67. https://doi.org/10.1098/rsta.2010.0327

Stockholm Resilience Centre. *Planetary Boundaries*. https://www.stockholmresilience.org/research/planetary-boundaries.html.

Tanner, Kathryn. *God and Creation: Tyranny or Empowerment?* Oxford: Blackwell, 1988.

Tilley, Terrence. *The Evils of Theodicy*. Washington, DC: Georgetown University Press, 1991.

Wallace-Wells, David. "The Uninhabitable Earth." *New York Magazine*, July 10, 2017. https://nymag.com/intelligencer/2017/07/climate-change-earth-too-hot-for-humans.html.

Whyte, Kyle. "Indigenous Climate Change Studies: Indigenizing Futures, Decolonizing the Anthropocene." *English Language Notes* 55:1 (2017) 153–62. https://doi.org/10.1215/00138282-55.1-2.153

Mosquitoes, Dengue, and Butterflies: Providential Masks of the Hidden God

Marisa Strizzi[1]

■ Our Share of Suffering

Astrophysics, quantum physics, and cosmology have taught us so far that the universe begins with the Big Bang. From this scientifically depicted event emerge different patterns of complexity: the succession of transitional moments from which something new abruptly appears and the whole system enters a new phase. Thus, the crossing of various thresholds takes us through the 13.7-billion-year history of our universe and the 4.5-billion-year evolution of the Earth. And there was the emergence of what we call "life" about 3.5 billion years ago with the evolution of the living, the development of our species—among millions of different others—through seven million years, and the more recent ten-thousand-year accelerated

---

1. Marisa Strizzi is professor and coordinator of the area of theology and interdisciplinary studies in the Ecumenical Network for Theological Education (REET), in Buenos Aires, Argentina. She is registered at the University of the Western Cape as a co-researcher for the project on "An Earthed Faith: Telling the Story amid the 'Anthropocene'."

> **How to cite:** Strizzi, M 2024, 'Mosquitoes, Dengue, and Butterflies: Providential Masks of the Hidden God', in EM Conradie & UL Vaai (eds.), *Making Room for the Story to Continue?*, An Earthed Faith: Telling the Story amid the "Anthropocene", vol. 4, AOSIS Books, Cape Town, pp. 181-198. https://doi.org/10.4102/aosis.2024.BK415.08

drama of human civilization. In effect, whatever has happened until now—independently of the narrative efforts to lay it out—greatly exceeds human agency. Human activity started to make a substantive difference, incipiently, ten thousand years ago against the background of an old and huge developing cosmos and within the delimited, extremely thin surface of one planet in one galaxy.[2]

We have arrived so incredibly recently on the scene. And "we" here stands for as wide a formulation as "homo sapiens" as possible, or, alternatively, as a punctual, minuscule designation as "theologians of the twenty-first century"—now making efforts to ponder the topic of providence against the background of the evils of our age. Who knows what would happen to any of us if we were able to quantify the amount and degree of suffering in a single second on Earth and display the intricacies of that measurement? Yet I dare to think that the majority of such suffering would result not from the vicissitudes of the expansion of the universe or of biological evolution but from the evil that humans inflict on one another and on the rest of the living and nonliving.

The Sum of All Evils

Images and stories about global warming and the related multifarious events that accompany them impose a threat. Our increasing knowledge about the "Anthropocene"[3] and its impact is a constant source of anguish and anxieties, for, in effect, avoiding the shock has become impossible, because it is not only a question of varied and disastrous meteorological changes, but of global events of a systemic nature, of geo-socio-economic-political-biological and cultural dynamics that entangle all of us. To speak of the impacts of the "Anthropocene" is to speak at the same time of a crisis of civilization, and this belongs in a scenario that appears as the sum of all evils.[4]

We inhabit a planet that suffers the impact of the hyper-realization and spreading of a modern Western civilizational pattern, marked by its character of anthropocentric, patriarchal, colonialist, classist, racist, and cultural homogenization. The world thus constructed bears the mark of the imposition of the human species over others and—within the human sphere—by the primacy of privileged subjects on the basis of sex, gender,

---

2. For a broader elaboration on these aspects, see Christian, *Maps of Time: An Introduction to Big History*; also the resources provided by the Big History Project, https://bhp-public.oerproject.com/ [last accessed October 11, 2024].

3. For the use of this term in this series, see the editorial introduction by Conradie and Vaai in this volume.

4. I have discussed aspects of this section more extensively in Strizzi, "Theological Affirmations for Living Together," 171-74.

eugenics, ethnicity, and class. The type of relations and exchanges that these subjects establish with their "others" are mostly in the form of asymmetrical, hierarchical bonds, which always involve some kind of violence. A peculiar aspect of this civilizational pattern is the expansion of an economic model that pursues endless growth, a fact that brings with it uncontrolled exploitation and systemic aggression of the elements that make life's subsistence possible. In reality, on planet Earth, the eco-*logical* has been subjugated by the eco-*nomic*, and this law (*nomos*) is unhinged: it obeys and imposes the dictates of capital. Thus, the evolutionary miracle of human knowledge and technoscientific development is manipulated and used within a logic co-opted by the law of progressive domination, which seeks to reduce everything it touches into merchandise and bring the profits "home." It will never be repetitive to state that the "home" of this *oikos-nomos* is not an open home that welcomes and shelters everyone. Rather, it is a transnational globalized construct that profits according to the scheme of privileges already observed, which perversely distributes the benefits on the one side and the consequences of obtaining them on the other. However, the management of the planetary consequences has long since escaped from the hands of this *oikos-nomos*.

The Evils at Home

I am writing this essay in a country where, at the present time, the year-on-year inflation rate is 254.2 percent,[5] where poverty rates are around 57.4 percent for the whole population and 64 percent in the case of children.[6] This disastrous panorama is part of a complex historical process that has worsened in recent years because of serious political and economic mismanagement, aggravated by an uninformed management of the quarantine imposed during the coronavirus disease 2019 (COVID-19) pandemic and a significant number of climatic phenomena related to global warming.

In terms of how climate change is affecting Argentina, data indicate that average temperatures are increasing in all latitudes and that some areas suffer great water stress, while others are experiencing more frequent extreme precipitation with subsequent floods. Conversely, the Andes Mountains have already shown signs of glacier loss.[7] At this very moment, some provinces are severely flooded because of the rising rivers resulting

---

5. See Instituto Nacional de Estadísticas y Censos de la República Argentina, "Índice de precios al consumidor (IPC). Enero de 2024," 3.

6. See Observatorio de la Deuda Social Argentina, "Informe: Argentina (2004–2023)," 29.

7. See, Barros, "Climate Change in Argentina"; Camilloni, "Argentina y el cambio climático."

from the heavy rains associated with El Niño, which has recently started. Other provinces are coming from a deep drought, having lost their harvests as a result of La Niña that predominated last year. Our news often includes desperate images of people evacuated because of fire or floods; farmers severely injured or who have lost their lives trying to save their crops or livestock; large tracts of territory affected and their flora and fauna devastated. Losses from untimely frost, floods, droughts, and fires are catastrophic for a country's economy that is mostly reliant on agriculture and cattle, and most dramatically for its people, who are already greatly impacted by political g/local economic factors. In addition to these climatic calamities, other associated evils of the "Anthropocene" play their part, environmental pollution being one of the most important. Different regions of the country are affected by the use of pesticides and other toxic agrochemicals, making workers and the nearby population sick. In the same way, mining ventures without strict governmental monitoring are not complying with the legislation, contaminating the soil and water streams.[8]

As I write, heavy rains and sustained heat, unusual for this time of year, have caused the incidence of dengue fever—a disease that occurs in tropical climates—to increase by 89 percent in the country. In Buenos Aires, the city where I live, long lines of infected people wait in hospitals to be treated. The breeding of the *Aedes aegypti* mosquito—the vector of this infection—has been largely driven by regional climate change. At the same time, the city is overrun with butterflies, something that hardly happens any longer. The heavy rains of the last few days were ideal for cocoons and chrysalides to complete their cycle. Just as millions of mosquitoes that infect us were born, the humid conditions also helped the butterflies that brighten us and pollinate our flowers to flourish. Mosquitoes, dengue, and butterflies: our unexpected realities.

Thus, within the context of the "Anthropocene" on planet Earth, in the first quarter of the twenty-first century and living in a large and rather doomed South American country—quite peripheral to Empire—I propose to contribute my small grain of sand to this conversation. I choose to do it by visiting the writings of a theologian who inhabited the outskirts of a declining Empire some five hundred years ago. For the fact is that while awareness about the "Anthropocene" and its evils started long after Martin Luther's time, the suffering of God's creation is not new, and this Reformer's theology provides a multifaceted approach to this issue and its implications. Actually, pondering some of Luther's distinctive motifs can contribute distinct perceptions to the present search. It is a truth that

---

8. See Aparicio, "Pesticide Pollution in Argentine Drinking Water"; Herrera, "Water Contamination at Barrick's Veladero Mine Threatens Health and Human Rights"; and Chivers, "Agricultural Pesticide Use in Argentina."

every generation needs to face its own challenges, yet we are always, unavoidably, inheritors of those who lived before us. Reading and rereading is part of our theological task.

■ A Theological Fix

This volume brings together different efforts that seek to answer a theological question formulated as: "How could the suffering of God's creatures in the so-called 'Anthropocene' be reconciled with trust in God's loving care?" My experience in the sphere of faith communities and theological education is that such a question is rarely put. There are certain key themes that people do express frequently when considering the sufferings related to the "Anthropocene." Firstly, they refer to a new global complex problematic. Secondly, they detect that the responsibility for this problematic is of human origin and, mostly, lies in the societies of the so-called "first world." Thirdly, the perception of injustice in this context is directly related to the previous and rarely to God. Questions about God's providence amid this situation seem to be absent, rather eclipsed by the call to denounce the responsibilities of transnational economic politics for their planetary excesses and the local establishment for their corrupt and inefficient management of such issues. Bible studies, sermons, thematic workshops, and informal conversations mostly reflect an absence of recourse to God's providence in this regard. Thus, the immediate context of this essay seems to indicate that we are here trying to answer a question that few are asking.

There are two primary elements that hardly appear together in theological discussions about the planetary crisis: divine action and human action. Actually, in this scenario, the scale mostly tips towards the latter. In a few words, general views are that if we emphasize human responsibility and agency, with all of its complexity, we will be successful in inspiring people to take up activism in an effort to reverse the myriad of problems associated with the "Anthropocene." Conversely, if we emphasize divine action and providence, people will become disengaged from the terrifying realities of the moment and adopt either a passive trusting attitude or a cynical stand. In effect, in secularized contexts, where cultures and theologies rely on the action and free will of the modern Western rational subject, clinging to divine providence becomes insulting. At best, the dilemma of human suffering and that of all creation is a thematic package dispatched to the land of theodicy and without return.

So the question that is not asked does relate somehow to the theological issue of God's providence. But it seems that common theological judgments are so unbalanced in this sense that the very notion arouses suspicion and heated reactions. Does the introduction of providence into this picture necessarily mean the exclusion of human responsibility and agency? Or is it that this bare option is considered an unbearable display of naivety? My stance is that reversing this scheme may actually lead to a shift that allows us to focus on the relationship between divine and human action in a broader, more nuanced and creative way.

A Providential Stance

Behind this essay lies the uneasiness of a noticeable "God-avoiding" attitude in ecclesial and theological communities engaged in processes of awareness and action in the face of the "Anthropocene" crisis. In the face of this, a question that pops up is whether some Christian faith communities and institutions are actually losing their faith, because there are different approaches—more or less systematic—to the theme of God's providence, and the introduction to this volume bears witness to that.[9] However, there is an aspect that is missing or underplayed in this discussion: this is that, unlike theodicy, a Christian understanding of providence does not belong to the field of rational questions and answers; providence pertains to the territory of faith.[10] The Christian trust in God's providence does not arise from a rational calculus of prediction but is elicited from promises formulated in Scripture. In the face of evil and suffering, believers may cry, lament, and complain only because a promise about justice and love has already been made and they have already believed it. Oswald Bayer well observes that, as humans, we do share with other living creatures in the inarticulate shrieks of pain, but human lament and cries of indignation before unjust suffering are possible because we know about love and justice through a word that has already been spoken to us.[11]

Indeed, the question of suffering in the "Anthropocene" puts believers in a difficult situation that has to do with faith. As Christians, we trust in God's providence, but in real life, our perceptions tell us that it isn't working. In effect, as Luther remarks to Erasmus, if we regard God's actions in the world following the judgment of human reason, we are forced to say "either that there is no God or that God is unjust."[12] Here the problem is not only that God acts in mysterious ways but that those ways are in contradiction to the given word of God. This is the experience of the hiddenness of God that Luther elaborates upon in his theology, and it belongs to a scenario in which God appears "inaccessibly distant and obtrusively close at the

---

9. See the essay by Conradie and Vaai, "In God We Trust?," in this volume.

10. This premise coincides with Gijsbert van den Brink's conclusion; see Van den Brink, "Does God Care for Oxen?" in this volume.

11. See Bayer, *Living by Faith*, 71–73.

12. See Luther, *The Bondage of the Will*, 215.

same time."¹³ Lest we confuse providence and theodicy, let me make it clear from the beginning that the latter has no place in Luther's theology. For Luther, it is God who works all in all, and God does not need human justification.

Thus, the premises that frame my readings on this regard are the following: firstly, that within the corpus of Martin Luther's works, the theme of God's providence is not a discrete, isolated topic but is woven into the whole of theology; secondly, that the issue of creaturely suffering is abundantly approached in this theology; thirdly, that Luther's theological motif of the hiddenness of God is a fruitful tool for exploring these themes.

I intend to follow the thread of some of Luther's elaborations in this regard, extricating some of its implications. This brief exercise may contribute useful insights to the present discussion. Given that more than five hundred years separate us from the author, this is necessarily an exercise in rereading.

■ God's Hiddenness

The recourse to the hiddenness of God comes up in some responses to the question of suffering in general and that of suffering in the context of the "Anthropocene" in particular. Be it a discussion about theodicy or providence, this appeal appears as a last resort and, in some cases, results in a somewhat abstract or vague formulation. Yet Luther's theology, in spite of its motley references about the issue, is quite clear and goes deep into the substance of matters. This is because he finds that knowledge of God only discloses itself to us indirectly, through mediations. Very early in his career, Luther abandoned any kind of speculative theology, and for this reason, the motif of the hiddenness of God does not refer to the absence of God or to a subject of metaphysical "being and not being" (as he argued in discussing Dionysian mystical theology);¹⁴ it refers to God's hidden, constant, and scandalous presence in creation.

Truly, Thou Art a God Who Hidest Thyself¹⁵

As Robert Jenson observes, this hidden presence of God is problematic for us humans because of its "offensive availability in our world."¹⁶ So much so that Luther uses the expression *"absconditus sub contrario"* to

---

13. See Bayer, *Living by Faith*, 70.

14. See Luther, *LW 14*, 110–11.

15. Parts of this section are taken from Strizzi, "Luther after Derrida."

16. See Jenson, *Systematic Theology*, 233.

speak about it.[17] God's revelation happens "hidden under its opposite"—at least, under what human morals and reason would consider as the opposite of divine revelation.[18] The very example of this is the cross. Although he has explored and elaborated on this topic during the time of his theological breakthrough, it is in the Heidelberg disputation (1518) that Luther introduces his "theology of the cross" confronting a "theology of glory."[19] He contends that the way of analogy—ascending from nature in search of the universal truth of God through abstraction—does not make a theologian; a theologian is the person who remains below, not climbing out of but staying in the muddy ground of reality. Because of this, a theologian of the cross does not speculate about the invisible things of God by applying reason to the visible traces in nature but perceives God's revelation in the *"posteriora Dei"*—God's rearward parts. For Luther, when one tries to perceive the invisible things of God through the created things, one just finds "virtue, godliness, wisdom, justice, goodness, and so forth," yet those findings are in opposition to God's "human nature, weakness, and foolishness" by which God wishes to be recognized.[20]

In fact, Luther discovers that creation does not refer to God in a symmetric analogical way but only by asymmetric paradox, as expressed by Vítor Westhelle: "It is asymmetrical in that what appears to be the case in one set of categories is not simply reflected in the other, but is shaped in the other in *unexpected ways*."[21] God hides in God's works, and those works operate differently from what human beings think they have unraveled. Reading Paul in 1 Corinthians 1:21, Luther concludes that it is not good for anyone to recognize God in God's glory and majesty, unless they know God in folly: the humility and shame of the cross. And, in this context, he links two scriptural truths: "God destroys the wisdom of the wise, as Isa. [45:15] says, 'Truly, thou art a God who hidest thyself'."[22]

Moreover, Luther finds that these theological aspects do not only refer to knowledge but also to morals. Epistemology and ethics are related: the way human beings know goes together with the way they act. Those who do not know the God hidden in suffering "hate the cross and suffering, and love

---

17. See Luther, *The Bondage of the Will*, 101.

18. This is discussed more extensively in Strizzi, *Luther after Derrida*, 69–80.

19. See Luther, *LW 31*, 40.

20. See Luther, *LW 31*, 52–53.

21. Westhelle *Transfiguring Luther*, 126 (emphases added).

22. Luther, *LW 31*, 53.

works and the glory of works";[23] they take credit for works and wisdom, but this wisdom turns out to be "completely puffed up, blinded, and hardened."[24] It is for this reason that "[a] theologian of glory calls evil good and good evil. A theologian of the cross *calls the thing what it actually is*."[25]

▪ God Unbound

In Luther's discussion with Erasmus years later,[26] the topic surfaces again, and what is argued there is the source of many theological disputes. The argument expands its semantic field, and the "God that hides" is now named the "hidden God." Luther explains that the *Deus absconditus* "neither deplores nor takes away death, but works life and death, and all in all."[27] The hidden God is "God unbound," God not limited by God's word. For such a reason, the hidden God does many things and wills many things that God does not show in God's word. Beyond these affirmations, Luther is concise and wraps up the issue quite sharply: we have nothing to do with the hidden God, for God's hiddenness is not a matter of speculation. Believers have to do with God "as clothed and displayed in God's word, by which God presents Godself to us."[28] The fact is that, reacting to the treatise on free will in which Erasmus favored a philosophical search while adducing the obscurity of Scripture, Luther affirmed God made known to us in Scripture and not in philosophical speculation.[29]

Luther should have known that asking readers not to lucubrate on the hidden God was asking something like "try not to think of a white bear." For when the hiddenness of God becomes food for speculation, one enters an unending loop. Yet having experienced himself the anguish of the "no way out," he does not avoid the issue; he explains that the very incarnate God weeps, laments, and groans before the hidden will of God![30] As for believers, this common experience of God's hiddenness in the world is so contradictory that they need "to find refuge in God against God."[31]

---

23. See Luther, *LW 31*, 53.

24. Luther, *LW 31*, 40–41.

25. Luther, *LW 31*, 40 (emphasis added).

26. Luther responds to Erasmus's *De libero arbitrio* (1524) with his *De servo arbitrio* (1525).

27. Luther, *The Bondage of the Will*, 170.

28. See Luther, *The Bondage of the Will*, 170.

29. This topic is discussed more extensively and in a more nuanced way in Strizzi, *Luther after Derrida*, 105–33.

30. See Luther, *The Bondage of the Will*, 176.

31. "… ad Deum contra Deum confugere," see Luther, *Operationes in Psalmos*, WA 5:204, 26–27.

The motif of the *Deus absconditus* appears as the fruit of human awe in front of what is unbearable and incomprehensible, but hardly as the source of faith. Still, as Luther also expresses to Erasmus, "[…] faith's object is things not seen. That there may be room for faith, therefore, all that is believed must be hidden. Yet it is not hidden more deeply than under a contrary appearance of sight, sense and experience."[32] The outrageous way of God's actions is a source of rational and moral rejection, and for this reason, "the highest degree of faith is to believe that God is merciful." Luther insists that we have to know God as incarnate, as Jesus crucified, "in whom are hidden all the treasures of wisdom and knowledge."[33] It is the word of the cross that bounds the God unbound.

In Luther's theology, that God is hidden *sub contrario* is part of the Christian paradox of recognizing God in the world; it refers to the cognitive as well as to the moral, but the former happens exclusively because of the latter. Yet it is not an apophatic option or a mechanism of escape in the face of suffering.[34] That God is hidden in shocking realities leads to the promises of God's word, anchored in the faith of Jesus Christ. This compels us to speak clearly and not linger in an unending loop of denials.

■ The Suffering of Creation

As Christian theologians, there is a detail that we cannot ignore: we inhabit a fallen reality. Every effort of our "task" is inscribed in a postlapsarian condition. For this very reason, human sinfulness is an unavoidable fact in our theological considerations. An experience that accompanies postlapsarian life is that of suffering, and this is directly associated with human sin. The Judeo-Christian account of the myth of the fall describes how a state of perfect peace and harmony becomes one of constant toiling and suffering. Indeed, suffering appears as an immediate punishment for sin, and throughout history, many elaborations have exploited—and continue exploiting—this vein of "crime and punishment." Different disciplines and sciences agree on the catastrophic impact of the species *homo sapiens* on planet Earth. They also record how this mark has intensified since a certain period in the history of civilization, so that the very name *anthropos* is the one chosen to identify this new epoch of the "Anthropocene." It is undeniable that the negative effects of the human species on planet Earth involves much suffering, and the complex web of this impact can be considered

---

32. See Luther, The Bondage of the Will, 101.

33. See Luther, The Bondage of the Will, 101.

34. *Pace* the editorial introduction by Conradie and Vaai to this volume.

theologically as sinful. Indeed, it is so in Luther's view: "since humans have fallen in sin, we all—the whole creation—must suffer the consequence."[35]

In his early *Lectures on Romans* (1515-1516), Luther offers insightful comments in relation to human beings and the suffering of creation. His notes, glosses, and scholia represent a turning point in his theology, two years before the disputation in Heidelberg.

The Nature of Humans

Reading Romans 8:3, Luther notes the inability of human nature to do good.[36] Confronting the thought of Aristotle and scholasticism, which sustains that human reason is inclined to seek the best, Luther does not deny this—for in effect, reason advises us to strive for the best. Instead, the question he poses is this: what is this "best"? Evidently, it is not what is good for others, because human nature in its knowing and desiring searches for what is good for itself; that is, it looks for what is good in an evil way.

Inspired by Scripture, Luther thus describes the picture of us human beings as "turned in" on ourselves. This happens in such a way that we use everything, not only what is physical but also what is spiritual, for our own purposes; in all things we only seek ourselves. This "curvedness" is characteristic of human sinful nature, which in its turning to itself bypasses everything else—even God— without seeing it. This is for Luther the very mechanics of idolatry. Human nature is for itself the first and greatest of all idols; consequently, it makes of God an idol; and finally, it makes idols of all created things and of all the gifts of God. An idol is a construction that answers our own desires, turning all others into what we need them to be. Conversely, it is grace that puts God in the place of everything else it sees, and also in its own place: it prefers God to itself and seeks only what is God's and not its own. Luther's theology proclaims that it is only through absolute grace that we can see God in and above all we see.[37]

Humans and Creation

Luther, like Paul, sees creation (in its postlapsarian state) as subjected to service. In his gloss to the text of Romans 8:19-22, Luther refers to the unwilling subjection to vanity of the creature and maintains that if one understands "creature" to refer only to "human beings," it is actually better to apply the term "vanity" to human beings only. It is because of this

---

35. Luther, *Sermons 8*, 90.

36. See Luther, *LW 25*, 344-45.

37. See Luther, *LW 25*, 346.

vanity—which is a wrong appreciation and misuse by humans—that creation becomes bad and harmful.[38] He links this to the interest of philosophers and metaphysicians in creatures, which he finds useless for theologians, announcing: "the time has come for us to give ourselves to other studies, and to learn Jesus Christ, 'and him crucified'."[39] Two years later, he will describe what goes on in this confrontation between glory and cross. Here, he already remarks that while we take pleasure and glory in our knowledge of the created world, creation mourns over its sufferings. It is not surprising that Paul—a theologian of the cross—discovers what Luther calls "an unusual and admirable theological term" when he discusses creation. Instead of looking for essences, operations, and states, Paul speaks of the "expectation" of the creatures. Actually, the apostle "hears" creatures expecting, hoping. Luther sums up: "Whoever studies the 'essences' and 'operations' of the creatures rather than their sighs and expectations, is without a doubt a blind fool. For such a one does not even realize that the creatures are creatures."[40] As Westhelle, reading Luther, expresses: "Only faith can see creation."[41]

Twenty years later, Luther creatively elaborates on this passage in the Epistle to the Romans in two sermons.[42] He voices what he had found in his early lectures and offers a very clear picture. He observes the suffering of all creatures—living and nonliving, human and not human—and also how human beings are directly implicated in such suffering. Remarkably, he clearly differentiates the suffering of believers as a consequence of their Christian witness in the world from the innocent suffering of all creation.[43] Besides, he notices that Christians—as humans—have an articulated language and resort to prayer, yet nonhuman creatures have speech, which is an accusing cry about their unjust abuse, and this is "intelligible to God and the Holy Spirit." Now, this accusing cry is "beyond human power to express, for God's created things are innumerable."[44] In effect, creation will be liberated as children of God, because "the creature in bondage has the same hope of release as the poor, enslaved human being."[45] Both Paul and

---

38. See Luther, *LW 25*, 362.

39. Luther, *LW 25*, 360.

40. Luther, *LW 25*, 361-62.

41. See Westhelle *Transfiguring Luther*, 126.

42. See Luther, *Sermons 8*, 82-100.

43. See Luther, *Sermons 8*, 91-94.

44. Luther, *Sermons 8*, 89.

45. Luther, *Sermons 8*, 98.

Luther see that, in spite of all human dominion, before God, creation is beyond human mediation and control.

■ The Hidden Workshop of God

In the context of the "Anthropocene," we are here posing the question about the suffering of God's *creatures*. And unless the use of the terms "creation" and "creature" has become an empty use of language, our concern and agency regarding God's creatures always imply a confession of faith. The symbol of our Christian faith starts out precisely by confessing our belief in God, the creator. In this sense, Gustaf Wingren points out sharply: "The concept of nature belongs primarily to dogmatics, not to ethics."[46] It is God, not us, who is at work constantly in the workshop of nature. And, in effect, God is *creating*. What the work of God demands "is not first action, but *faith*."[47] Nature becomes creation by faith; God's providence belongs to the same territory.

Lecturing on Genesis, Luther affirms that for God, "creating and preserving are identical."[48] Preaching on the Gospel of John, he assures us that God not only creates the world but works constantly in it and through it, because God is "with all creatures, flowing and pouring into them, filling all things."[49] Elaborating on the first article of the creed, he explains what this confession means: in the first place, it means that God has created us along with all creatures;[50] in the second, it means that none of us has our life or anything else that has been created in our own hands, nor can we preserve any of them, however small and unimportant.[51] For Luther, the God who creates is the God who sustains all creation. And confessing God as the Creator is confessing the Triune God. In fact, Luther finds that the "mystery of the Trinity" can be read in the narrative of Genesis. So he explains: "The Father creates heaven and earth out of nothing through the Son, whom Moses calls the Word. Over these the Holy Spirit broods [...] For it is the office of the Holy Spirit to make alive."[52]

---

46. Wingren, "The Doctrine of Creation," 364.

47. Wingren, "The Doctrine of Creation," 364.

48. Luther, *LW* 4, 136.

49. Luther, *LW*, 22:26.

50. As the German version of the Small Catechism states: "*Jch gleube, das mich Gott geschaffen hat sampt allen Creaturn.*" See Luther, *WA 30.I*, 363, ¶2.

51. See Luther, "The Large Catechism," II, 15–16.

52. Luther, *LW* 1, 9.

The Masks of God

Also reading Genesis, Luther affirms, "God is the one who is hidden. This is God's peculiar property."[53] The hiddenness of God permeates Luther's theology and is inherent to what believers confess. God is closer to us than we are to ourselves,[54] for in reality, "nothing can be more truly present and within all creatures than Godself and God's power."[55] And, since we cannot see God directly, Luther depicts God hidden in God's words, in God's works, in all creation. In relation to this, in many of his writings, Luther uses the peculiar motif of the *larvae Dei*—the "masks of God." The *Lectures on Galatians* offer a good example: there Luther speaks of the whole creation as the mask of God; for instance, vocations, offices, and social roles are each a mask of God.[56] The hidden God works behind the mask of human work. In an exposition of Psalm 127, he asserts that "the course of the world" is God's mask.[57] As Larry Rasmussen observes, in Luther, "we only know the creator, redeemer, and sustainer in finite disguises [...] hidden behind one mask or another."[58] Still, whether in the carnival—medieval or postmodern—or in everyday life, masks are tricky, for "we should know that God hides Godself under the form of the worst devil."[59] As already observed, God *absconditus sub contrario* is a call for faith and not for analogical games.

God is always present, working all in all, and creation is the mask of God's mediated immediacy. The hidden God in Luther is the constant promise of God's transcendent presence in immanence. Since we are not the creators or the sustainers, we must admit that there are aspects in the workshop of creation that exceed us, and not only a few of them. Providence is faith arising in what we confess every time that we confess God. The hidden God is no other God than the Triune God, creator, sustainer, justifying savior, and life-giver, and our agency is part of God's hidden, masked work.

Living in the World

So far, we have put a lot on God's side of the balance. Luther is an expert at that. Now, where does human agency fit into this? If God is the one who works all, hidden in all things, what do *we* do? Reading Luther, perhaps the

---

53. Luther, *LW* 6, 148.

54. See Luther, *LW* 37, 60.

55. Luther, *LW* 37, 38.

56. See Luther, *LW* 26, 95–96.

57. See Luther, *LW* 45, 331.

58. Rasmussen, *Earth Community, Earth Ethics*, 279.

59. Luther, *LW* 7, 175.

answer is that we can do nothing to save God's creation and we can do absolutely everything to take care of our neighbor, that is, of all those who have been created alongside us. In effect, for Luther, the life of believers is the continuous paradox of living before God and before the world at the very same time. Perhaps one of the best examples of this is his well-known formulation: "A Christian is a perfectly free sovereign of all, subject to none. A Christian is a perfectly dutiful servant of all, subject to all."[60] This *simul* is understood as the key of Christian freedom. We could say that because God receives and saves by absolute grace, believers have a clear, free conscience before others in everything they do; because others are in need, believers are called to serve others freely, seeking the good of all others rather than their own. Believers are to love the world and their neighbors. For this reason, we are rational beings, we develop scientific knowledge, we confront social injustice, we fight for better laws, and we act responsibly, fulfilling our vocations and working on behalf of others. And Luther says that we should do it "as if there were no God."[61]

Creation is a matter of faith which goes together with grace. The world is the place we inhabit, and God´s providence is the tissue that holds us back from falling into the abyss of distorted, self-righteous explanation of ourselves and of the world. Scientific data tell us about complex aspects of what goes on in cosmic reality; they are a correlate which accompanies and informs our theological questions in and about the world. From there, we get a map to guide our Christian actions, actions that we set forth trusting that God is at work at it, because we have a call to serve our neighbor. Theology knows about reality in a postlapsarian condition. Here we experience the imperfection of things and the structural limitations of our agency. So suffering tells us different stories about a simultaneous event. In one of them, we are human beings that act responsibly with and for others; in the other, we discern our human sinfulness, and we hear the moaning and perceive the expectation of all creatures. We suffer because we are alive; we suffer because we inhabit a fallen reality. In Luther's theology, suffering is not a form of achievement; suffering is just a fact that we have to face.

In spite of all his late medieval imagery and historical constrictions, in Luther, what anchors our trust in God's providence is not a proposal of *credo quia absurdum*. It is the faith that holds us when we learn Christ and him crucified. Luther's Christology is the heart of his theology, and it implies a radical reading of the Christian *credo*. For it is there that the materiality of God's presence in the world takes momentum. For Luther, God cannot

---

60. See Luther, *LW* 31, 344.

61. See Luther, *LW* 45, 331.

be God without this crucified, suffering human being, and this human Christ is unconditionally present everywhere, where the right hand of God is.[62] As we confess the Christian creed and believe in the triune God—whose love creates, justifies, and saves, opening the future—we are engaged in a reality that we cannot see in what we see.

■ Providential Masks

As we search "providential" answers under the influx of the "Anthropocene," Luther's readings may tell us that knowing Jesus Christ, and him crucified, remains still the way of doing theology. This implies to refrain from romanticizing the cross once again. The crucified shows us what the nature of our species is. Through the ages, human herds evolved, settled, and grew based on force and violence; through the centuries, we continued cementing civilizations that are culturally based on the exaltation of dominion, the trampling of the weak, the worship of beauty, the ostentation of curved goodness and the repetition of the same. Then the innocent tortured creature, brought to trial and executed on the cross, tells us who we are and where God is. Westhelle's perception allows us to revisit the hidden God and the masks, so drenched in Luther's late medieval imagination. The Brazilian theologian writes:

> Auschwitz, Hiroshima, or the hole in the ozone layer are the masks of God raised for our self-recognition, in which we measure ourselves as much as in the lilies in the fields. The nature we see is the mirror image of what we have made it, or have allowed it, to become.[63]

Luther may teach us to perceive the believers' contradictory existence in the world and our need of finding refuge in God against God. This is actually a depiction of our struggles facing the evils of the age, and we know that the picture is not a smiling selfie. Luther's hard, realistic view of human beings in the world may sober us up. The cross means that there is no glory in Christian life. Let there be no doubt: to "learn Christ and him crucified" embraces the triumph of resurrection, but the grammar of this formulation goes hand in hand with the theologies of those who have been vanquished so many times. Such is the faith of the risen one, who is none other than the crucified one.[64]

---

62. This topic is addressed in the conclusion to the editorial introduction to this volume. See Conradie and Vaai, "In God we Trust?" The reference there is to Karl Barth's treatment of the doctrine of God's providence. Luther's theology is different in this regard and the discussion on this difference is not new.

63. See Westhelle, *Transfiguring Luther*, 129.

64. See Sobrino, *Jesús en América Latina,* 235.

Here and Now

Providence is a matter of faith, and as Christians we need to persist in it. After all, this is our way of being in the world, that is, by "the certainty of what is hoped for, the conviction of what is not seen." But faith, far from being a form of naïve resignation or a resource of the intimate gymnastics of the soul, is openness to the other that exceeds us. The faith of Jesus Christ—the justice of the triune God—puts the human in context and examines it; it dismantles the dazzling fantasies of the ego, the pretensions of dominion of the rational subject, the structures of slavery, the sin of the species. Faith is the only thing that allows us to participate in another reality, one that captivates us by its pneumatic promise, that continually opens to the future. Thus, it interrupts us and removes us from the suffocating search for "our own," from the curvature, from navel-gazing—not only as individuals and communities, but also as a species. To bear witness to this faith is to point out the evils and to name them, persisting untiredly in the hopeful practices that grace grants us, because grace manifests itself in impossible scenarios where the injustice of the *nomos* is confronted by what cannot be commodified. It is only in the face of the rotting and suffering fabric of so many masks that the gift occurs. Relationality and care, respect of the other—human and nonhuman, living and nonliving—and the renunciation of violence as a resource are the works of grace in and through us. We are provoked to discern the hidden, masked God at work in the world—constantly and outrageously loving.

Bibliography

Aparicio, Victoria, and Eduardo De Gerónimo. "Pesticide Pollution in Argentine Drinking Water: A Call to Ensure Safe Access." *Environmental Challenges* 14 (2024) 1–10. https://doi.org/10.1016/j.envc.2023.100808

Barros, Ricardo, et al. "Climate Change in Argentina: Trends, Projections, Impacts and Adaptation." *WIREs Climate Change* 6:2 (2014) 151–69. https://doi.org/10.1002/wcc.316

Bayer, Oswald. *Living by Faith. Justification and Sanctification.* Translated by Geoffrey W. Bromiley. Grand Rapids: Eerdmans, 2003.

Chivers, Charlotte. "Agricultural Pesticide Use in Argentina: The Extent, the Risks, and the Challenges." 1 February 2022, *SPRINT Sustainable Plant Protection Transition*. https://sprint-h2020.eu/index.php/blog/item/6-pesticides-argentina.

Christian, David. *Maps of Time: An Introduction to Big History*. Berkeley: University of California Press, 2004.

Camilloni, Inés. "Argentina y el cambio climático." *Ciencia e Investigación* 68:5 (2020) 5–10.

Herrera, Viviana, and Jan Morrill. "Water Contamination at Barrick's Veladero Mine Threatens Health and Human Rights." *Minning Watch Canada*, 2 December 2022. https://miningwatch.ca/blog/2022/12/2/water-contamination-barrick-s-veladero-mine-threatens-health-and-human-rights.

Instituto Nacional de Estadísticas y Censos de la República Argentina. "Índice de precios al consumidor (IPC). Enero de 2024." *Índice de precios* 8:5 (2024) 1–16.

Jenson, Robert W. "The Hidden and Triune God." *International Journal of Systematic Theology* 2:1 (2000) 5–12. https://doi.org/10.1111/1463-1652.00024

———. *Systematic Theology I: The Triune God*. Oxford: Oxford University Press, 1997.

Luther, Martin. D. *The Bondage of the Will. A New Translation of the De servo arbitrio (1525), Martin Luther's Reply to Erasmus of Rotterdam*. Translated by J. I. Packer and O. R. Johnston. London: James Clarke, 1957.

———. "The Large Catechism." In *The Book of Concord: The Confessions of the Evangelical Lutheran Church*, edited and translated by Theodore G. Tappert, 357–461. Philadelphia: Fortress Press, 1959.

———. *Luther's Works*, edited by Jaroslav Pelikan, Helmit Lehman and Christopher Boyd Brown. Saint Louis: Concordia, 1955- (incomplete).

———. *Martin Luthers Werke. Kritische Gesamtausgabe*. Weimar: Hermann Böhlaus Nachfolger, 1883–2009.

———. *Sermons by Martin Luther*, edited and translated by Nicholas Lenker. Grand Rapids: Baker, 2000.

Observatorio de la Deuda Social Argentina. "Informe: Argentina (2004-2023). Un régimen inflacionario crónico de empobrecimiento y mayor asistencia social." *UCA-Pontificia Universidad Católica Argentina*. https://uca.edu.ar/es/noticias/presentacion-informe-avance-deudas-sociales-cronicas-y-desigualdades-crecientes.

Rasmussen, Larry. *Earth Community, Earth Ethics*. Maryknoll: Orbis Books, 1996.

Sobrino, Jon. *Jesús en América Latina. Su significado para la fe y la cristología*. Santander: Sal Terrae, 1982.

Strizzi, Marisa. *Luther after Derrida. The Deconstructive Drive of Theology*. Lanham: Lexington /Fortress Academic, 2022.

———. "Theological Affirmations for Living Together." In *International Handbook for Creation Care and Eco-Diakonia*, edited by Dietrich Werner et al., 169–77. Oxford: Regnum Books, 2022.

Westhelle, Vítor. *Transfiguring Luther. The Planetary Promise of Luther's Theology*. Eugene: Wipf and Stock, 2016.

Wingren, Gustaff. "The Doctrine of Creation. Not an Appendix but the First Article." *Word and World* 4:4 (1984) 353–71.

God's Providence and Suffering from Climate Change in the African Context

Gloriose Umuziranenge[1] & Eraste Rukera[2]

■ Introduction[3]

The task before the churches in Rwanda in view of climate change and the emerging condition of the "Anthropocene" is daunting. The scope of the crisis is, of course, planetary. We have come to know that climate change causes catastrophes and disasters which kill millions of people, and this

---

1. Eraste Rukera is the administrative assistant of the dean of the Faculty of Theology at the Protestant Institute of Arts and Social Sciences (PIASS) in Rwanda and a minister in the Presbyterian Church in Rwanda. He is registered as a co-researcher for the project on "An Earthed Faith: Telling the Story amid the 'Anthropocene'."

2. Gloriose Umuziranenge is a senior lecturer and the director of quality assurance at PIASS. She is registered as a co-researcher for the project on "An Earthed Faith: Telling the Story amid the 'Anthropocene'."

3. We wish to acknowledge the assistance of Clive Pearson in editing this essay (EMC & ULV).

How to cite: Umuziranenge, G & Rukera, E 2024, 'God's Providence and Suffering from Climate Change in the African Context', in EM Conradie & UL Vaai (eds.), *Making Room for the Story to Continue?*, An Earthed Faith: Telling the Story amid the "Anthropocene", vol. 4, AOSIS Books, Cape Town, pp. 199–214. https://doi.org/10.4102/aosis.2024.BK415.09

makes climate change a challenge to our time.⁴ In a similar vein, António Guterres, the Secretary-General of the United Nations, observed that "[m]ore severe and frequent floods, droughts and tropical storms, dangerous heatwaves and rising sea levels are severely threatening lives and livelihood across the planet."⁵ In a recent statement, Guterres declared that "humanity has opened the gates of hell" and the prospect of a "dangerous and unstable future beckons."⁶ This warning is far from being isolated and idiosyncratic. A number of writers have expressed the seriousness of climate change by indicating that the viewpoint of "irreversible climate change has convinced many people that the survival of societies, cultures and of the planet itself is at risk."⁷ An ecumenical conference on the present being a *kairos* for creation, held at Wuppertal in 2019, concluded that "the urgency of the situation implies that a comprehensive response cannot be delayed."⁸ The level of concern and sense of urgency has only intensified as one year follows another.

This planetary crisis expresses itself in different ways from one context to another. In terms of an interdisciplinary task of reading the implications of such for Rwanda, the specific types and levels of threat are laid out in the national reports to/of the Intergovernmental Panel on Climate Change (IPCC), the World Bank, and the national Environment Management Authority.⁹ The formal nature of such reporting should not obscure the ordinary, real-time experience of extreme events and the demands it places on a nation's capacity for resilience and its infrastructure. The writing of this chapter comes in the wake of heavy rain and floods that killed people in the Western Province and destroyed much infrastructure and many plantations. Those who managed to escape were put into refugee camps.

The particularity of the situation which presents itself to the churches in Rwanda is reflected in a series of statements made by denominations, singly and collectively. They can be set alongside other ecumenical initiatives further afield, most notably those from the World Council of Churches (WCC) and the Orthodox and other African churches. The tendency of such statements is to fasten upon practical steps, ethical

---

4. PIASS, "Together for Survival"; Eglise Anglicane du Rwanda (EAR) and Eglise Presbyterienne au Rwanda (EPR), "One for Climate," 2–3.

5. World Meteorological Organization, "Statement," 4.

6. Milman, "Humanity," ¶1.

7. United Nations Environment Plan, "Environment Religion and Culture," 5; Subramaniam, "Crisis of Consumption," 8.

8. Andrianos et al., *Kairos for Creation*, 12.

9. IPCC, "East Africa"; World Bank, "Rwanda"; Rwanda Environment Management Authority (REMA), "State of Environment, 2021."

exhortations, and immediate hopes, justified by a selected biblical text or theme. The intention is the admirable one of seeking to address the question of what churches can do to reduce the Earth's climate crisis. The most notable of those responses come from the Protestant Council of Rwanda and the initiatives of Anglican Diocese of Shyogwe. From a theological perspective, the position adopted is often one of seeking to present creation as God's property and, not surprisingly, how the work of stewardship or guardianship might be seen in an eschatological perspective.

In this present essay, the line of enquiry is altered: the issue of climate change, the "Anthropocene," and theology are addressed via the doctrine of providence and a question that is often left implicit in this *kairos* situation. The sometimes-close relationship between providence and theodicy is worked out in a Rwandan response to the question "how could the suffering of God's creatures in the 'Anthropocene' be reconciled with trust in God's loving care?" It is likely that this framing of the question is not necessarily the one that presents itself to those working in the field. It is not surprising that in the light of the relatively low levels of greenhouse gases produced and emitted in Africa, the sequence of questions can take other forms: Why does Africa continue to be a victim even though it has contributed less to the problem? Does God care for Africa as God does for other continents? Does God care for his creatures over the world in the same way? Does God not take care of them because they are called the "sons of Ham"? What these questions signify is how the reading of providence in this context is readily linked to a theological posing of what constitutes climate justice. They bear the resonances of a theology of rejection. The theodicy question here becomes one of "why are they suffering if God is caring for them?"

This topic is not an easy one for a theology emerging from Rwanda. That is so for two reasons. The first has to do with the concept of the "Anthropocene," which is not widely known. It belongs to a sphere that is removed from everyday life and ordinary understanding. In times gone by, it was possible to invoke the tradition of rainmakers. Now, the context has changed through the knowledge represented by the climate sciences. Praying for rain is a rather limited ecclesial response. This way of viewing the state of the planet and interpreting environmental change through the lens of climate science lies beyond Indigenous forms of knowledge: it is not to be found in the worship and life of discipleship in the local churches. It presupposes complex levels of interdisciplinary research. Rwanda in general is a country of oral tradition: the telling of stories functions as a means of building a strong bridge between present and future generations. It is designed to address negative behavior and reinforce good practice and desirable communal values. It is a very different kind of epistemology, the purpose of which in the present is whether such storytelling is able to speak to the issues of a changing climate and (maybe) introduce talk of the "Anthropocene." Can storytelling help bridge the gap between those who may have the technical expertise of climate science and theology and those for whom these types of knowledge are alien?

■ Making a Statement

Whether formal church statements on the state of the planet ever directly address the critical theological questions is an open question. They tend to describe a problem and plot a course of action; they are then inclined to exhortation. They can nevertheless carry an implied set of theological convictions and suggest where further work might be done.

The initiatives taken by churches in Rwanda do so against a background of such faith traditions. As the effects of climate change became more dangerous, the Christian faith has become more aware of the threats posed and the need for a practical theological response. In 1989, the Ecumenical Patriarchate drew attention to the duty of the church to use all the theological and spiritual means at its disposal for the protection of planet. Earlier still, in 1983 at its Vancouver Assembly, the WCC urged member churches to commit to a platform of justice, peace, and the integrity of creation. At its tenth assembly in Busan in 2013, it defined the victims of climate change as a new kind of poor, alongside widows and strangers who need care and the love of God. Meeting at Shanghai and Nanjing, China (November 17–23, 2016), its Executive Committee adopted the "Statement on Climate Justice," in which it declared that countries that go beyond national interests should take responsibility to safeguard those suffering from effects of climate change. The necessity of changing lifestyles to address the root causes of climate injustice was assumed. The eleventh assembly held at Karlsruhe, Germany (from August 31 to September 8, 2022) was organized around the theme "Christ's Love Moves the World to Reconciliation and Unity." The World Student Christian Federation (WSCF) participated. The clear message released was a call to the world to take action in the following short sentence: "the time to act is now."[10] The World Council has sent delegations participating in all Conferences of the Parties (COPs) seeking "to remind the negotiators that beyond the technical aspects of mitigation, adaptation, technology transfer and funding, there is an ethical responsibility which ought not to be overlooked."[11]

It is now reasonably common for one Protestant denomination after another to show a concern for planetary well-being in the face of climate change.

---

10. WCC, WSCF at the 11th General Assembly of the WCC, 8.

11. Rautenbach et al., *Religions for Climate Justice*, 19.

The same is true of the Roman Catholic Church, most notably evident in Pope Francis's encyclical *Laudato Si'* (2015) and the apostolic exhortation *Laudate Deum* (2023). These documents build upon previous initiatives from the Vatican. Such initiatives included the visit of Pope Paul VI to the Food and Agriculture Organization of the United Nations (FAO), on the 25th anniversary of its institution. On that occasion, he warned the world that a progressive deterioration of the environment would provoke a veritable ecological catastrophe if measures were not taken and put into practice.[12] In 1971, Pope Paul VI wrote the encyclical *Pacem Terris*, in which he demonstrated that "ecological concerns" are an anthropogenic "tragic consequence."[13]

This deepening global ecumenical concern is further reflected in the specific focus on ecological degradation and the risks associated with the climate crisis arising out of African churches. Of particular interest is the work of the South African Council of Churches. It has set out a platform of six actions. The first is committed to the introduction of an earthkeeping program within the church service. The second action relates to the establishment of eco-congregations. The third is to initiate various earthkeeping projects within the life of the church. The fourth is a commitment to advocacy and fighting against economic injustice. The fifth is the desire to include lay involvement in the overall platform. The last of these six actions fastens upon the adoption and promotion of ways of living in Christian families that do not conflict with a positive climate change agenda.[14] It is recognized that these steps are part of a necessary global campaign.

This practice of issuing ecumenical statements is also to be found in Rwanda. In this particular instance is the work of the Conseil Protestant du Rwanda (CPR) or Protestant Council of Rwanda. Its initiatives are not well documented, but three statements stand out. The first is the "Theological Charter of the Environment" (2011).[15] It lays the theological and practical foundation for all subsequent praxis. Strictly speaking, the Charter is not explicitly theological: it is not so in a sustained manner. It is more a statement of values, calls, and commitments into which biblical references are inserted. They are confined to the following: Genesis 1:26, 28; 2:7; 2:15; 3:19; Colossians 1:15-20; and Romans 8:28. The Charter is nevertheless understood "as a sign of our gratitude to God the Creator, the Provider and Sustainer of Life."

---

12. Pope Paul VI, Visit of Pope Paul VI to the FAO on the 25th Anniversary of Its Institution, 3.

13. Pope Francis, *Laudato Si'*, 4.

14. SACC, *Climate Change—A Challenge to the Churches in South Africa*, 16-17.

15. CPR, "Theological Charter."

It is on this basis that the Charter concerns itself with the need to mobilize people for environmental care in four successive sections. The opening three comprise the response of people in general, followed by work at a family and a community level. The present is a time to be an "advocate for the environment" and being aware that "we will have to give an account for how we have fulfilled the mission of protecting it."[16] The importance of aligning the churches' strategic plan with the National Environment Plan is noted. It presumes supporting and promoting the Rwandan values on environmental care being in accord with the gospel.[17] This fusion of faith and context lends itself to conflicting interests. In keeping with this aim, "trained pastors" are called to draw upon Rwandan culture "in aspects of poems, songs, dances, proverbs, tales […] and from the Bible."[18] The last section of this Charter reinforces the imperative of humanity's responsibility for the well-being and guardianship of the environment. Failure to be mindful of this vocation runs the risk of the human species destroying itself.[19]

The second document is entitled "African Commitments for a Living Planet" (2012).[20] It was designed to be a "response to the aggressive deforestation in the country from 1990 to 2003 following war and genocide."[21] The word "commitment" here is critical. It signifies a number of potential outcomes that are designed to fulfill the aspirations of the Charter. They are listed below:

- fifteen seminars that should be conducted per year with a target of training 10,500 pastors, youth, and women in the church on degradation of environment and climate change
- the establishment of fifteen tree nurseries, including fruit tree nurseries
- the planting of one million trees per year, in open spaces, starting with church land (each parish should plant ten thousand trees in three years)
- the distribution of one million tree seeds
- the production of an environmental charter adhered to by parishes
- the establishment of a task force to monitor the project of climate change
- the composition and making available of an explanatory narrative and financial report.

---

16. CPR, "Theological Charter," Commitments 3 and 4.

17. CPR, "Theological Charter," Commitment 8.

18. CPR, "Theological Charter," Commitment 10.

19. CPR, "Theological Charter," Commitment 12.

20. CPR, "African Faith Commitments."

21. CPR, "Theological Charter," 1.

The task of achieving this plan was placed in the care of the CPR. That aim sits alongside two complementary projects—the Food Security Training Program and Campus Green.

The third document is an article by Gloriose Umuziranenge on the theme of "The Role of the Churches in Addressing Climate Change."[22] Rwanda serves as a case study through which Umuziranenge puts the Charter and other initiatives into perspective. She explains how Rwanda is a "highly populated and landlocked country" where "the potential of agricultural production is relatively high."[23] The present dilemma facing the nation is not simply a result of the direct influence of climate change, which has seen the emergence of broken weather patterns marked by droughts and flooding. Umuziranenge describes how urbanization is pushing "farmers into increasingly fragile lands" which has then resulted in habitat loss and the degradation and pollution of soil and water.[24] The emphasis on the planting of trees becomes readily understood in the light of war and genocide against the Tutsi, along with other violent conflicts in neighboring countries and the hosting of refugees. The impact of conflict on the land is intensified by the reliance of the people of Rwanda on "trees as a source of energy." Umuziranenge notes that 96 percent of households are dependent upon this form of energy and charcoal for cooking, hence the emphasis placed on afforestation and the family in the document devoted to a living planet. Umuziranenge identifies the need for a holistic program that will also include "the promotion of energy saving technologies, i.e. stoves and solar lighting for a clean environment."[25]

■ The Practice of Resilience

The task before the churches in Rwanda is daunting. As is the case in many other parts of Africa, there is low community awareness and a lack of resources. In their study on the Global South, Anna Piggott-McKellar and associates surveyed the project and donor reports (the grey literature); they did so in order to identify the barriers (rather than limits) facing community-based adaptation projects in those parts of the Global South that have a "heightened vulnerability" to the impact of climate change.[26] Three types of barrier were identified: the sociopolitical systems in which

---

22. Umuziranenge, "Role of the Churches," 209.

23. Umuziranenge, "Role of the Churches," 209.

24. Umuziranenge, "Role of the Churches," 211.

25. Umuziranenge, "Role of the Churches," 211.

26. Piggott-McKellar et al., "What Are the Barriers?" 375.

organizations like the church must operate;[27] those to do with resources, which include the pressure of time, a lack of information, and access to technology, alongside limited financial and human resources; and the need to overcome the physical and process barriers that make it difficult to cope with the actual impact of natural hazards.

Umuziranenge cites a workshop held at Nyamata and a project run by the Shyogwe Diocese of the Anglican Church in order to highlight the gap between intention and fulfilment. The churches in Rwanda "still face the challenge of low community awareness about the risks of climate change which constitutes an obstacle to ownership of protection measures, lack of knowledge and skills among the community and church members concerning climate protection, and difficult collaboration between environmental actors (churches, civil society, foreign agencies, and government services and lack of enough resources to implement projects to address climate change."[28]

Similar observations were made at the Nyamata workshop on "One for Climate" (2016). Rather tellingly, the report indicates that there is an absence of "climate justice and environmental protection on the church agenda and in its all programs, a lack of cooperation with other churches and institutions in addressing climate change, lack of a network of churches to promote climate justice and environment protection."[29]

The possibility of resilience has nevertheless been reflected in the experience of the Anglican Church of Rwanda (EAR) Shyogwe Diocese. The vulnerability assessment report of the Rwanda Environment Management Authority (REMA) shows that the Southern Province is the most vulnerable to climate change effects in Rwanda. However, the same report indicates that the district of Muhanga, where the headquarters of Shyogwe Diocese are located, has the lowest vulnerability among the eight districts of the Southern Province because of the combination of relatively low impact value and high adaptive capacity.[30] The initiatives and responses of EAR Shyogwe Diocese to climate change are many: they include energy-saving stoves, disaster risk reduction and environment protection, carbon mission reduction, and the use of ceramic water filters. The diocesan buildings observe a protocol for sorting waste through the 4Rs of reduce, reuse, recycle, and recover before it reaches the landfill.[31]

---

27. Piggott-McKellar et al., "What Are the Barriers?" 382–83.

28. Umuziranenge, "The Role of the Churches," 323.

29. Rural Development Interdiocesan Service (RDIS), "Workshop on the Role of Churches in Addressing Climate Change."

30. REMA, "Assessment of Climate Change Vulnerability," 17–20.

31. RDIS, *Waste Management for Environmental Protection in Rwanda*, 2.

■ Establishing a Theological Framework

How are these and other initiatives now to be set within a compelling theological narrative? How is that to be done in a way that resonates with the lived experience of this particular context? The Rwandan experience suggests that the abrupt rupture into the "Anthropocene" does not necessarily stretch the theological response to climate change to a breaking point (as might be the case elsewhere). That is not necessarily a problem for how an attempt is then made to reflect on the idea of providence and shape a response to related concerns of theodicy on behalf of suffering creatures. The very idea of providence lies in the intersection between what is universal and what is particular.

How a concern for God's care and a desire to express an opinion on what is God up to in this challenging time may take different forms from one location to another. In the case of Rwanda, the three-fold pattern, expressed elsewhere in this volume, of *conservatio*, *concursus*, and *gubernatio* may not be formally employed, but that does not mean that there is no understanding of God as provider and sustainer. It is then aided and supported by the biblical call to stewardship.

While the concept of providence itself may not often be spoken about, belief in the providence of God is fundamental in the African context. John Mbiti maintains that "[i]n various ways, God provides for things he has made, so that their existence can be maintained and contained."[32] The underlying conviction is that God provides all that is needed for sustaining God's beloved creation. God is known among different African peoples as the provider of sun, rain, and water, considered as blessings. This implies that God takes care of the whole creation and human beings do not have to worry about life (Matt 6:25–34). Life itself is indeed a gift from God—for the Banyarwanda, God is *Rurema*, which means Creator in our mother tongue (Kinyarwanda). Writing on Indigenous African theologies, Gabriel Boitshepo Ndhlovu presumes that divine providence should be "understood as the active role of God in the affairs of this world." There are "four views" or aspects to this claim: "God causes all things, God directs all things, God controls by liberating, and God limits such control."[33]

There is a prior step to be taken, though, given the role of Scripture in a Christian reading of providence and climate change. The very idea of providence presumes a desirable end, the fulfillment of God's purposes. That expectation is placed alongside the ending of one era—the Holocene—and the potential beginning of another—the "Anthropocene"—which

---

32. Mbiti, *African Religions and Philosophy*, 41.

33. Ndhlovu, "Reshaping South African Indigenous Theology," 82.

attracts discourses about the extinction of species and an "ugly" future for humankind. The prospects of a compassionate or benevolent understanding of providence seem to be radically compromised. It is a scenario that is capable of releasing a rather grim eschatological perspective through texts like Isaiah 51:6 and 2 Peter 3:5-13. The former text envisages the Earth wearing out like a garment; those who live on it will die like gnats and the heavens will vanish like smoke. The epistle imagines that on the day of the Lord, the sky will disappear; the material of elements of the universe will be destroyed by fire, and all that humanity has constructed will be destroyed. This type of eschatological passage is found in many other places (Ps 102:25-27, Mark 13:31, Heb 1:10-12). Starvation, diseases, and earthquakes are the signs that indicate that the end of this world is near (Luke 21:11). In a setting where the impact of climate change can be extreme—like in Rwanda—some care must be exercised lest the idea of God's providence lead to the people believing that such suffering is an instrument of divine pedagogy or punishment.

The importance of this proviso cannot be ignored. Umuziranenge notes that Rwanda is overwhelmingly Christian, with 95 percent of the population so classified. Its people are "deeply imbedded [sic] into religion systems and rationalities." In this kind of context, she argues that is a "cultural language which people in Rwanda understand well."[34] It matters, then, what biblical narratives are privileged and how some others, potentially harmful or indifferent to the environment, are managed.

One example of such a risk can be seen in the light of annual climate statements that now feature in the emerging "Anthropocene": they often assume a time scale of when a tipping point may be surpassed and conservation becomes difficult with the loss of biodiversity—and, maybe, the prospect of mass extinctions and endings. The Earth might then be regarded as a wasteland of sorrow, exile, and environmental calamities. It is not difficult to see how, in such circumstances, a raft of biblical texts might then be invoked in order to place all hope on an interventionist God or a yearning for heaven at the expense of caring for creation. There is then no need to act wisely.

It would seem as if the practical measures being adopted in places like the Shyogwe Diocese most readily lend themselves to readings of *conservatio* and *concursus*. That this should be the case is suggested by the call made at the Nyamata conference. Bishop Kalimba Jered turned to Genesis 2:15 in order to lay claim to the churches being stewards of God's creation. It is the churches' responsibility to take care of all creatures,

---

34. Umuziranenge, "Eco-Theology and Climate Justice," 265.

including human beings, animals, and nature, which is our common home.³⁵ On several occasions Bishop Karimba, in advocating climate action, refers to the desire for a "New Eden."³⁶ At a follow-up conference to the "One for Climate" workshop (2018), those participating were invited to strive for "the reconciliation with [...] creation by bringing back the Garden of Eden" and to work together in order "to be blessings to our nations and transform the lives of peoples."³⁷

This turn to Genesis 2:15 is found elsewhere in African writings. Benjamin Diara and George Christian Nche from Nigeria argue that this Scripture "clearly echoes man's [sic] responsibility to his environment." It is a requirement of God and, in their way of thinking, places "a repeated emphasis on the goodness of creation and on humankind's role in working for and with God in the process of 'repairing' creation."³⁸

There is nevertheless a potential ambiguity in this turn to Genesis 2:15: it presupposes the keeping or care of the garden, as well as tilling or working it for the sake of ongoing life. It is one thing for an individual farmer to clear his land of weeds, rocks, and tree stumps; it is of an altogether different order for humankind to dominate rainforests and destroy fertile soils with massive machinery. It is the scale of dominion that is now the problem.³⁹

The risks inherent in this second purpose are intensified when the two creation stories in the opening chapters of Genesis are run together. The notion and practice of care may then be overpowered by Genesis 1:28, with its mandate to multiply, exercise dominion, and subdue. The apparent right to dominion can easily become one of domination. At a conference held in Kibuye, Benoit Gerardin noted how important it was to move away from readings of domination and exploitation, which belonged to the different setting of the people of Israel facing foreign powers and captivity.⁴⁰ In the words of the Ethiopian Orthodox Tewahedo Church's Archbishop Abba Aregawi, creation must not be considered as mere raw material but as a gift to humans.⁴¹

---

35. EAR and EPR, Workshop on "One for the Climate," 4.

36. See RDIS, http://rdis.org.rw/workshop-11-2018 [last accessed October 11, 2024].

37. RDIS, http://rdis.org.rw/workshop-11-2018 [last accessed October 11, 2024].

38. Diara and Christian, "Theology of Climate Change," 85.

39. Rukera, "Assessing the Role of Churches in Mitigating and Adapting Climate Change," 21.

40. Gerardin, "Creation Care," 270.

41. Nzwili, "African Faith Leaders."

■ Proverbial Sayings

The standard practice in offering a theological response to climate change in a variety of workshop and conference statements is to invoke a handful of biblical texts. What lies a little out of view to an outsider is the role of distinctive Rwandan ideas and values. Sometimes they are expressed through proverbial sayings like those associated with children. One such saying is "*Imana yirirwa ahandi igataha i Rwanda*" ("God passes the day away, and in the night, he comes to Rwanda"). In the evening, families are always required to have water in their houses so that when God comes, there is water for the sake of enlarging the family. This water is called "God's water." The sustaining role of God in family life is also seen in the names given to children, for example, Hatangimana ("God is the Giver") and Harerimana ("God takes care of our children"). God creates and sustains.

The way in which the providence of God does not rule out the need for human beings to act and collaborate is evident in another proverb. The following saying, "*uragiriwe n'Imana ashyiraho n'uwe mushumba*," can be translated as "even if the cattle are cared for by God, the owner has to hire his own cowboy." It expresses as such an idea of responsibility. The presumption that one of the benefits of providence might well be the blessing of prosperity is countered by another proverbial saying—in this instance, "*Imana iraguha ntimugura, iyo muguze iraguhenda*," meaning that God gives his blessing for free. It cannot be bought. In the light of this saying, the practice of a prosperity gospel cannot be justified in terms of the providence of God.

■ Conclusion

Belief in the providence of God does not imply human passivity in the face of the environmental threats to the well-being of God's creation. The extremes of weather are making themselves felt in Rwanda. Their impact on vulnerable lives and the natural order itself can lead to the cries of theodicy, reflections on the nature of sin, and invocations of climate justice. The language of the "Anthropocene" is remote to most people in Rwanda. They are not alone in that situation. Their knowing of the "Anthropocene" is through experience. In this case, the experience is not just a consequence of global warming trends: it also arises out of the aftermath of war and genocide, where there has been a significant loss of trees and agricultural capacity.

Faith and culture provide a framework that professes that creation is a gift of a God who cares and exercises the three-fold roles of *conservatio*, *concursus*, and *gubernatio*. These words in this form may not feature in popular expressions of faith or the Indigenous knowledge of the

Banyarwanda peoples. The same mingling of worldviews lends itself to the practical need to care for creation and build resilience in the face of suffering and the powers of sinful exploitation of the Earth's resources.

It is only right that human beings take the responsibility of caring for other creatures and the preservation of the "common home." It might also be argued that the suffering of creation and innocent peoples intensifies an aspect of the belief in providence that is seldom mentioned. The suffering on the cross is an illustration of God's solidarity with his people in suffering. The destruction of God's creation and its suffering are interconnected and go together with God's suffering. God suffers with and for his creation. Failure to act in keeping with the providential care of God in a time of climate change causes God to suffer.

The sheer complexity of climate change and the very nature of ecological boundaries require multidisciplinary action. The solution does not lie simply in extraordinary technical achievements and further economic growth, however desirable that may be in some quarters. There is an urgent need for an accompanying ethic and transformation of life, to which the providence of God can testify. Greed and the unrestrained desire for unlimited economic growth have pushed human beings to overexploit the creation and inflict suffering on God's people. To sustain the Earth for future generations, a radical transformation is required.

■ Bibliography

Andrianos, Louk, et al., eds. *Kairos for Creation: Confessing Hope for the Earth*. Solingen: Foedus-verlag, 2019.

Bloomquist, Karen L. *God, Creation and Climate Change: Spiritual and Ethical Perspectives*. Geneva: LWF, 2009.

Chianeque, Luciano. "Creation Theology and Ethics of Sustainability." In *African Christian Theologies and the Impact of the Reformation*, edited by Heinrich Bedford-Strohm et al., 307-14. Zürich: LIT Verlag, 2017.

Church of England. *Climate Change: The Policy of the National Investing Bodies of the Church of England and the Advisory Paper of The Ethical Investment Advisory Group of The Church of England*. 2015. https://www.oikoumene.org/sites/default/files/Document/climate.change.policy.30.04.15.pdf.

Conradie, Ernst M. "Praying for Rain? Reformed Perspectives from the Southern African Context." *The Ecumenical Review* 69:3 (2017) 315-26. https://doi.org/10.1111/erev.12295

CPR. *Role of Church Parishes in Protecting, Rehabilitating the Environment and Fighting against Climate Change in Rwanda*. Kigali: CPR, no date.

Diara, Benjamin, and George C. Nche. "Theology of Climate Change Mitigation and Adaptation: The Place of the Church." *Academic Journal of Interdisciplinary Studies* 2:13 (2013) 85-91. https://doi.org/10.5901/ajis.2013.v2n13p85

Davies, Geoff, and Kate Davies. "Priorities for Ecumenical Eco-Ethics in the 21st Century: Thoughts and Lessons from SAFCEI, a Multi-Faith Eco-Justice Organisation in Southern Africa." In *Kairos for Creation: Confessing Hope for the Earth. The "Wuppertal Call,"* edited by Lukas Andrianos et al., 281-89. Solingen: Foedus-verlag, 2019.

Eglise Anglicane au Rwanda and Eglise Presbyterienne au Rwanda. "Workshop on 'One for the Climate'." Hosted by EPR and EAR Churches with an exposure visit to RDIS's Stove project in EAR Church, Shyogwe Diocese in Rwanda, on 29th September, 30th September, and 1st October 2016, Kigali.

Francis (Pope). *Laudato Si'. Encyclical Letter on Caring for our Common Home.* Vatican: Vatican Press, 2015.

Gerardin, Benoit. "Creation, Care and Sustainability: Ethical Challenges." In *African Christian Theologies and the Impact of the Reformation*, edited by Heinrich Bedford-Strohm et al., 263–92, Zürich: LIT Verlag, 2016.

IPCC. "The IPCC's Sixth Assessment Report: Impacts, Adaptation Options and Investment Areas for a Climate Resilient East Africa." https://cdkn.org/sites/default/files/2022-09/IPCC%20Regional%20Factsheet_East%20Africa_WEB.pdf.

Kambale, Jean Bosco, and Bwiruka Kahongya. "Eco-Theology in an African Perspective: Why the Delay to Embody Eco-Theology in African Christianity?" In *Kairos for Creation: Confessing Hope for the Earth. The "Wuppertal Call*,*"* edited by Lukas Andrianos et al., 121–31. Solingen: Foedus-verlag, 2019.

Karamaga, André. *L'Evangile en Afrique: Ruptures et Continuité.* Yens/Morges: Edition Cabédita, 1990.

Kirkpatrick-Jung, Anna, and Tanya Riches. "Towards East Asian Ecotheologies of Climate Crisis." *Religions* 11:341 (2020) 1–22.

Lugazia, Faith K. "The Holy Spirit, Eco-Justice, and an African Lutheran Response to Ecological Peril." *Dialog: A Journal of Theology* 55 (2016) 282–86. https://doi.org/10.1111/dial.12264

Mash, Rachel. "The Anglican Church of Southern Africa and Climate Justice." *Scottish Episcopal Institute Journal* 5:3 (2021) 21–28.

———. *Renewing the Life of the Earth: Christian Discipleship and Environmental Action.* Cambridge: Grove Books, 2021.

Mbiti, John. *African Religions and Philosophy.* London: Heinemann, 1969.

Milman, Oliver. "'Humanity Has Opened the Gates of Hell' by Letting Crisis Worsen, UN." *The Guardian*, 20 September 2023. https://www.theguardian.com/world/2023/sep/20/antonio-guterres-un-climate-summit-gates-hell.

Motte, John. *Protecting Climate and the Environment Together: Programmes and Activities in the UEM Member Churches of Africa and Asia.* Wuppertal: UEM, 2015.

Mpofu, Buhle. "Pursuing Fullness of Life through Harmony with Nature: Towards an African Response to Environmental Destruction and Climate Change in Southern Africa." *HTS Theological Studies* 77:4 (2021) 65–74. https://doi.org/10.4102/hts.v77i4.6574

Ndhlovu, Gabriel Boitshepo. "Reshaping South African Indigenous Theology on God and Sin: A Comparative Study of Augustine's *Confessions*." *Conspectus* 19 (2015) 79–103. https://doi.org/10.10520/EJC170479

Nzwili, Fredric. *African Faith Leaders Urged to Push for Tangible Climate Change Action.* 2021. https://www.oikoumene.org/news/african-faith-leaders-urged-to-push-for-tangible-climate-change-action.

Paul VI (Pope). *Visit of Pope Paul VI to the FAO on the 25th Anniversary of its Institution.* Vaticana: Libreria Editrice, 1970. https://www.vatican.va/content/paul-vi/en/speeches/1970/documents/hf_p-vi_spe_19701116_xxv-istituzione-fao.pdf.

Peracullo, Jeane C. "Justice as Cry of The Earth: The Book of Revelation in Shin Megami Tensei's Devil Survivor 2." *Journal of Dharma* 39:1 (2014) 85–106.

PIASS. "Together for the Survival of Our Planet: Our Contribution to Mitigate Global Warming and Climate Change." *Piass Scientific Week*, 2018. http://piass.ac.rw/2018/07/25/piass-scientific-week-2018.

Piggott-McKellar, Annah E., et al. "What Are the Barriers to Successful Community-Based Climate Change Adaptation? A Review of Grey Literature." *Local Environment* 24:4 (2019) 374–90. https://doi.org/10.1080/13549839.2019.1580688

Pillay, Miranda N. "The Church and the Environment: On Being Down to Earth in a Consumerist Era." *Scriptura* 107 (2011) 184–98.

Piper, John, and Justin Taylor, eds. *Suffering and the Sovereignty of God*. Wheaton: Crossway, 2006.

Protestant Council of Churches of Rwanda. *African Faith Commitments for a Living Planet*. Kigali: CPR, 2012. http://www.arcworld.org/downloads/Rwanda-CPR-Summary-Sep2012.pdf.

———. *Theological Charter of the Environment*. Kigali: CPR, 2011. http://www.arcworld.org/downloads/Rwanda-Theological-Charter.pdf.

Rahman, Muhammad Ishaq-ur. "Climate Change: A Theoretical Review." *Interdisciplinary Description of Complex Systems* 11:1 (2013) 113.

Rautenbach, Ignatius, Kerber Guillermo, and Christoph Stückelberger. *Religions for Climate Justice: International Interfaith Statements 2008–2014*. Geneva: Globethics.net, 2014.

RDIS. *Follow-Up Workshop on The Role Of Churches In Addressing Climate Change in Rwanda 11-2018*. With Opening Remarks and Morning Prayer by Bishop Jered Kalimba and Closing remarks by Bishop Abednego Keshomshahara, Shyogwe, November 2018. https://rdis.org.rw/workshop-11-2018.

RDIS. *Waste Management for Environmental Protection in Rwanda: "If Waste is Sorted at the Production Site, any Further Handling and/or Treatment is Made Easier and Cheaper!"* Muhanga: RDIS, 2018.

REMA. *Assessment of Climate Change Vulnerability in Rwanda—2018*. Kigali, 2019. https://rema.gov.rw/cc_vulnerability_Rwanda(2018)-Final_report.pdf.

———. *State of Environment and Outlook Report 2021: Summary for Policymakers*. www.rema.gov.rw/fileadmin/user_upload/Rwanda_SOER_-_Summary_for_Policy_Makers_Final-HR.pdf.

Rukera, Eraste. "Assessment of the Role of Churches in Mitigating and Adapting to Climate Change: Case Study of Anglican Diocese of Shyogwe in Rwanda." Master's thesis, PIASS, 2023.

Singer, Peter. *Practical Ethics: Second Edition*. Cambridge: Cambridge University Press, 1999.

South African Council of Churches. *Climate Change—A Challenge to the Churches in South Africa*. Marshalltown: SACC, 2009. https://www.smms.ac.za/wp-content/uploads/2019/08/Climate-Change-A-Challenge-to-the-churches-in-South-Africa.pdf.

Subramanian, Kalpathy Ramaiyer. "The Crisis of Consumption of Natural Resources." *International Journal of Recent Innovations in Academic Research* 2:4 (2018) 8–19.

Tsalampouni, Ekaterini. "Dealing with Eco-Justice from the Perspective of the Orthodox Tradition: Challenges and Dilemmas." In *Eco-Theology, Climate Justice and Food Security: Theological Education and Christian Leadership Development*, edited by Dietrich Werner and Elisabeth Jeglitzka, 237–62. Geneva: Globethics.net. https://www.globethics.net/documents/4289936/13403236/GE_Global_14_web.pdf/.

Umuziranenge, Gloriose. "Eco-Theology and Climate Justice in Rwanda." *Kairos for Creation: Confessing Hope for the Earth. The "Wuppertal Call,"* edited by Lukas Andrianos et al., 263–69. Solingen: Foedus-verlag, 2019.

———. "The Role of Churches in Addressing Climate Change: Case Study of Rwanda." In *Eco-Theology, Climate Justice and Food Security: Theological Education and Christian Leadership Development*, edited by Dietrich Werner and Elisabeth Jeglitzka, 209–14. Geneva. Globalethics.net, 2016.

UNEP. *Environment, Religion and Culture in the Context of the 2030 Agenda for Sustainable Development*. Nairobi: UNEP, 2016.

United Nations Network on Migration. *East African Community Climate Change Master Plan.* 2015.

Vischer, Lukas, ed. *Report and Papers from a Consultation on Creation Theology Organised by the European Christian Environmental Network at the John Knox International Reformed Center from March 28th to April 1st 2004.* Geneva: Centre International Reformed John Knox, 2004. https://www.lukasvischer.unibe.ch/pdf/2004_creation_groaning.pdf.

WCC. *Cooler Earth: Higher Benefits Actions by Those Who Care about Children, Climate and Finance.* 2nd ed. Geneva: WCC, 2021.

———. *Presbyterian Church of Trinidad and Tobago: "Instead of Just Activists, We Create Strategists."* https://www.oikoumene.org/news/presbyterian-church-of-trinidad-and-tobago-instead-of-just-activists-we-create-strategists.

———. "Statement on Climate Justice, World Council of Churches Executive Committee, Shanghai and Nanjing, China 17–23 November 2016". https://www.oikoumene.org/sites/default/files/Document/23%20REV%20Statement%20on%20Climate%20Justice.pdf.

———. "WSCF at The 11th General Assembly of the World Council of Churches Karlsruhe, Germany". https://www.wscf.ch/docs/programmes/global-collaborations/WSCF_at_the_11th_GA_of_the_World_Council_of_Churches.pdf.

WMO. "WMO Statement on the State of the Global Climate in 2019". https://library.wmo.int/doc_num.php?explnum_id=10211.

World Bank. *Rwanda: Climate Risk Country Profile, 2021.* https://climateknowledgeportal.worldbank.org/sites/default/files/2021-09/15970-WB_Rwanda%20Country%20Profile-WEB.pdf.

Zizioulas, John. "Pope Francis' Encyclical Laudato Si'." In *Eco-Theology, Climate Justice and Food Security: Theological Education and Christian Leadership Development*, edited by Dietrich Werner and Elisabeth Jeglitzka, 179–86. Geneva: Globethics.net. https://www.globethics.net/documents/4289936/13403236/GE_Global_14_web.pdf/.

Does God Care for Cows? A Dutch Perspective on the Perils and Prospects of Providence in the "Anthropocene"

Gijsbert van den Brink[1]

■ Introduction

In a much-discussed passage the apostle Paul suggests that, despite the obvious literal meaning of Deuteronomy 25:4 ("Do not muzzle an ox while it is treading out the grain," New Revised Standard Version, Updated Edition [NRSVUE]), God is not concerned for fair treatment of oxen or other animals. Instead, he interprets these words as allegorically expressing God's

---

1. Gijsbert van den Brink is professor of Theology and Science at the Vrije Universiteit Amsterdam and extraordinary researcher at the Unit for Reformational Theology and the Development of the South African Society, Faculty of Theology, North-West University, South Africa. As a white, male, Western European theologian in the Reformed tradition, I realize that my positionality may predispose me to favoring certain perspectives on providence while obscuring others.

How to cite: Van den Brink, G 2024, 'Does God Care for Cows? A Dutch Perspective on the Perils and Prospects of Providence in the "Anthropocene"', in EM Conradie & UL Vaai (eds.), *Making Room for the Story to Continue?*, An Earthed Faith: Telling the Story amid the "Anthropocene", vol. 4, AOSIS Books, Cape Town, pp. 215-234. https://doi.org/10.4102/aosis.2024.BK415.10

care for human beings—in particular for underpaid Christian missionaries like Paul himself. Some contemporary interpreters go to great lengths to argue that *this* text (1 Cor 9:9-10), in turn, should not be taken literally but may have a meaning that is not totally indifferent to the fate of animals.[2] Others, however, give short shrift to such attempts and reject Paul's words out of hand as morally insensitive. For example, in his groundbreaking work *Animal Liberation*, atheist philosopher Peter Singer argues that the New Testament lacks any interest in the well-being of animals, quoting Paul in this connection: "'Does God care for oxen?' Paul asks scornfully. No, he answered, the law was intended 'altogether for our sakes'."[3] Even Albert Schweitzer, himself a Christian, conceded: "Of course Paul's exegesis is here at fault. It is part of the greatness of the legislation of Deuteronomy, that in this and so many other ordinances it does imply that God concerns Himself about the animal creation."[4] Indeed, Christian ecotheologians have been keen to paint a more nuanced picture, highlighting important strands in the biblical literature that actually do affirm God's care for the animals and suggest a non-anthropocentric appreciation of their value.[5]

I do not intend to address (let alone solve) this particular debate here. I mention it in order to show that debates on the scope of God's care or providence can already be found in the normative sources of Christianity, and that it is not so easy to settle them. In this contribution, cows stand *pars pro toto* for the nonhuman creation at large, or rather for the entangled web of creation that humans are irreducibly part of, i.e., for the biosphere. Cows may symbolize the intricate connections between and within the world's ecosystems.[6] For example, for many centuries, humans have depended on cows as much as the other way around. And living in the Netherlands, I am familiar with the way in which intensive stock farming and the bio-industry surrounding it capitalize on domesticated animals, exploiting (especially) cows to raise huge incomes while causing disconcerting levels of methane and nitrogen emissions which threaten our

---

2. E.g., Gilmour, *Eden's Other Residents*, 33-36. Gilmour points to the wider context of Deuteronomy 25:4 and concludes: "[…] Paul's great insight in these verses [in 1 Cor 9] is that God is not concerned with oxen *only* in Deut. 25:4, any more than other laws demanding generosity […] are about human beings *only*" (35, italics in the original).

3. Singer, *Animal Liberation*, 209.

4. Albert Schweitzer, *Mysticism of Paul the Apostle*, 320 (as quoted by Gilmour, *Eden's Other Residents*, 32).

5. See, e.g., Linzey, *Christianity and the Rights of Animals* (characterizing Singer's overview as "vastly oversimplified," 163); Preece and Fraser, "The Status of Animals"; Bauckham, *Living with Other Creatures*, 1-13, 79-132; Gilmour, *Eden's Other Residents*, 26-55; Horrell et al., *Ecological Hermeneutics*, 13-120; Deane-Drummond, *Primer in Ecotheology*, 18-35; Clough, "The Bible and Animal Theology," 402-08; compare Van den Brink, *Reformed Theology*, 101-06.

6. An impressive account of this profound entanglement can be found in Wirzba, *This Sacred Life*, especially 63-122; Wirzba rightly criticizes the notion that "environments are essentially *separate* from people" (103).

ecosystems and thus our future—including, ironically, our agricultural future. Of course, the farming industry itself argues that its work is needed to "feed the mouths" of the ever-increasing world population, but it is clear that doing that in this particular way continues the longstanding human addiction to meat-eating and has a disastrous long-term impact on the world's ecosystems.

To what extent is this entangled web of creation, with all its creatures "great and small," the object of God's care and providence? And what does that mean now that we live in a time in which, because of human exploitation, this web is increasingly becoming fragile? In other words, what does God's providence mean in the "Anthropocene," here to be loosely defined as the era in which—as a result of patterns of mass production and consumption in the Global North especially—we are confronted with irreversible depletion of natural resources, massive biodiversity loss, the extinction of many species, global climate change, and other forms of ecological degradation? Can we trust God to take care of all this, as the one who started the entire project by creating the world in the first place and situating planet Earth in a life-sustaining cosmic environment? Or to put all this in this volume's lead question: how could the suffering of God's creatures in the so-called 'Anthropocene' be reconciled with trust in God's loving care?[7]

In this contribution (as in this volume), these questions will be focused on the notion of divine providence—traditionally the place where talk about God's care for creation and creatures finds its doctrinal home. Some suspect that the "Anthropocene" may offer the final blow to a doctrine that has already lost much of its credibility as a result of the cultural regime of modernity and the horrors of twentieth-century wars and genocides.[8] Others suggest, conversely, that the "Anthropocene" should prompt us to return to an Augustinian discourse of human sinfulness and divine grace "as mediated by [...] divine providence,"[9] or to retrieve deeply existential notions from the theology of Martin Luther.[10] In any case, there is ample reason to revisit this doctrine in light of the global ecological challenges we are facing today. In what follows, I will first highlight why providence has

---

7. For an illuminating discussion of the extent to which the "Anthropocene" "changes everything," and also for an account that addresses the lead question of this volume in a way that "stays with the trouble," see Elisabeth Pyne's essay in this volume.

8. Pearson, "God's Continued Providence," 399.

9. Sideris, "A Rupture in the Earth," 20. Sideris argues that in the "Anthropocene" we may need "a story closer to that of premoderns who saw themselves—perhaps rightly, after all—as threatened and sustained by an all-powerful supreme being" (34).

10. See Maria Strizzi's contribution to this volume.

become such a risky notion in this connection, in order to then zoom in theologically on what the doctrine intends to convey and how it is supposed to function. Finally, I will apply these insights to our current ecological predicament.

■ The Perils of Providence in the "Anthropocene"

In a cynical mood, one might be inclined to say that whatever take on God's providence with regard to animals, plants, minerals, and other parts of the nonhuman creation we prefer, its outworking is detrimental anyhow. For if God does not care for the cows (as Paul suggests), and if we can indeed generalize that view so as to include other nonhuman beings and systems, then why should *we* care about nonhuman nature? Apparently only we humans really count in God's eyes, and surely our own care for other beings need not go beyond God's care. On the other hand, if God's providence *does* imply care and concern for nonhuman beings, then we might draw a similar conclusion with regard to our own responsibility. Clearly, if God takes care of nonhuman species and ecosystems, they will be safe in the end anyhow, irrespective of our own dealings with them.

Of course, things are not as simple as that. As indicated above, it has become more and more clear over the past decades that the human fate is inextricably bound up with that of the biosphere at large. For example, we now realize almost collectively that the current global climate change is real and has a huge impact on humans. The same applies to the massive loss of biodiversity through species extinction.[11] Thus, even if we were to hold onto an anthropocentric perspective, we would still have ample reason to care about other species and ecosystems. Yet even though sheer climate denialism seems on the wane today, many people, including Christians, persist in an attitude of relative indifference with regard to the biosphere.[12] There is little doubt that one of the most prominent strands of thought that leads groups of Christians—and presumably other religious believers as well—to such an attitude is the notion of divine providence.[13] This especially comes into prominence in the debate on global climate change. A dominant

---

11. An insightful exploration and continuation of recent theological work on species extinction and its consequences can be found in the volume of articles edited by Eva van Urk-Coster, "Our Current Extinction Crisis."

12. Even on the right, climate denialism is decreasing, for obvious reasons; see, e.g., Milman, "Climate Denial is Waning on the Right." For a broader analysis, see Veldman, *The Gospel of Climate Skepticism*.

13. A second theological locus that is influential here is eschatology; e.g., escapist end-times scenarios also fuel climate indifferentism among Christians. I will not explore this particular discourse here. See also Gloriose Umuziranenge's essay in this volume.

take on divine providence in this debate is that (despite Paul's remark on oxen) God *does* care about the nonhuman dimensions of creation, including the climate, so that as a consequence we humans should not be too bothered about it. Instead, we should trust God.

Thus Alan White, discussing the possibility of global warming in a contribution to the website of the creationist institute *Answers in Genesis* (based in Petersburg, Kentucky in the United States) argues as follows: "To those who believe that the heavens and the earth were designed and created by a 'higher' power, there is ample reason to expect that earth's temperature will remain in a range to support life. In fact, God gives us that promise in Genesis 8:22: 'While the earth remains, seedtime and harvest, cold and heat, winter and summer, and day and night shall not cease'. Within this worldview it makes perfect sense that the earth would have a temperature control system just like our bodies do, since God designed them both."[14]

Here we find an unambiguous appeal to God's providential care for creation, used as an alibi for human indifference and inactivity in fighting global climate change. Those who think that such crude statements can only be found in popular writings should be cautious. Firstly, one should not underestimate the spread and influence of such popular writings.[15] It is not without reason that forms of "operant" and "espoused" theology have recently received more explicit attention in practical theological research.[16] In this way, similar appeals to divine providence as implying human passivity have even spread to Christian populations in parts of the majority world that are in immediate danger as a result of climate change.[17] Secondly, on closer scrutiny, such statements are often backed up by more formal and normative theological sources. For example, influential conservative theologian Wayne Grudem likewise argues that the Creator's continued care for creation warrants that human activities which rely on the Earth's resources will not backfire: "Activities that produce carbon dioxide—such as breathing, building a fire to cook or keep warm, driving a car or tractor, or burning coal to produce electricity [… are] morally good and necessary that God intended for us. It seems very unlikely to me that God would have

---

14. White, "Climate Change Facts," also featured in Ham and Hodge, *The New Answers Book 4*, chapter 16.

15. *Answers in Genesis* promotes its 2022 edition of *The New Answers Book 4* by mentioning that over 750,000 copies of the series are in print.

16. See Cameron et al., *Talking about God in Practice*, 53-56. Their four-fold division of theology in operant (the theology embedded in the practices of a group), espoused (theology drawn from a group's own belief articulations), normative (theology from sources that the group considers normative), and formal (academic theology) has found rapid recognition and adoption in the field of (practical) theology.

17. See, e.g., the testimonies to this effect mentioned by Seforosa Carroll and Upolu Vaai in their contributions to this volume.

set up the earth to work in such a way that these good and necessary activities would actually destroy the earth."[18]

Thus, the belief that God will incessantly exert providential care and control over nature's systems provides a license to continue unabated the Global North's present-day patterns of production and consumption, including the implied high levels of greenhouse gas emissions that propel climate change. Presumably, similar arguments can be adduced with regard to biodiversity loss and other ecological disasters: God will take care that at the end of the day they will not thwart us humans in our accumulation of wealth. To doubt this can even be considered a sign of unbelief. Indeed, in an official document of the so-called Cornwall Alliance it is argued that being strongly concerned about ecological developments such as the current climate crisis amounts to a lack of trusting and acknowledging God.[19] The appeal to God's providence in particular is manifest here: "God's wisdom, power, and faithfulness justify confidence that Earth's ecosystems are robust and will by God's providence accomplish His purposes for them."[20]

In my view, this testifies to what we might call the *perils* of providence, that is, the risks that are involved in appealing to notions of divine care and providence prematurely or otherwise inappropriately. As some recent observers have perceptively commented, appeals and arguments to the effect that human activities cannot lead to dramatic forms of climate change since "God is in control" mostly "[…] arise from highly privileged contexts out of touch with the great [ecological] harm and suffering that humans have perpetrated throughout history."[21] Obviously, such arguments contribute to the ongoing (and more and more irreversible) occurrence of such harms, not only to the earth's ecosystems but also to its most vulnerable people who suffer profoundly from the dire consequences of climate change.[22] Ironically, in terms of the question that motivates this volume ("How could the suffering of God's creatures in the 'Anthropocene' be reconciled with trust in God's loving care?"), we must conclude that trust in God's loving care can, as a matter of fact, *aggravate* the suffering of God's creatures in the "Anthropocene"! Apparently, therefore, trust in

---

18. Blunt, "Cool on Climate Change," 26–27; Grudem, *Politics According to the Bible*, 367.

19. Cornwall Alliance, "Renewed Call," 17: "Fear of environmental catastrophe grows out of lack of the fear of God," a sentence that is followed by an even more perplexing one: "That is the real root of the many false or exaggerated environmental scares that have plagued the modern world." The Cornwall Alliance has been named after the place in Connecticut where its founders first met in 1999.

20. Cornwall Alliance, "Renewed Call," 17.

21. Lowe et al., "Climate Skepticism," 433.

22. See, e.g., Pearson, "God's Continued Providence."

God's care or providence is not by definition a good thing. We should be more specific as to when it is appropriate and when, on the contrary, it is theologically misplaced.

The abuse of the appeal to divine providence is evident in all times. We find instantiations of it even in the Bible itself, for example, in Jeremiah 7, where Israel's political leaders wrongly trust that the temple of the Lord, which they had made into a "den of robbers" (7:11), will never be demolished since God will "of course" take care of it. Recent examples include the use of the doctrine for undergirding apartheid policies in twentieth-century South Africa[23] and the suggestion that God had provided for Aryan leadership by the Nazi regime in the Second World War. Now, the easiest solution to prevent such abuses may seem to jettison the doctrine of divine providence altogether. But that would be to throw the baby out with the bathwater, since the belief that God is actively involved in and provides for creatures belongs to the core convictions that are constitutive of the Christian faith.[24] Dropping it would mean to end up with a deist view of life. Moreover, the abuse of belief in God's providential care does not annul the possibility of making proper use of it. Thus, the question now becomes how we might distinguish between proper uses of the doctrine and irresponsible, exploitative ones. Addressing this question requires that we zoom in a bit more on the roots of the doctrine and its place in the overall scheme of Christian teaching.[25]

■ The Particularities of Providence

Traditionally, the doctrine of divine providence was known as an *articulus mixtus*, a "mixed article" of the Christian faith, in the sense that it can be based both on biblical grounds and on more general philosophical reflections.[26] Indeed, for many centuries, belief in God's care and control over earthly affairs was experienced as quite self-evident. As Hendrikus Berkhof has pointed out, belief in providence has always been one of the most popular parts of the Christian faith. For many people, the belief "that nothing happens by chance" even "seems to be the whole of their faith."[27]

---

23. See the discussion in Ernst Conradie's essay in this volume.

24. See Hoggard Creegan's contribution to this volume for some very helpful suggestions as to how the doctrine of divine providence can accommodate the contemporary evolutionary worldview without losing its integrity.

25. For a similar exploration of the problems and prospects of doctrines of election and predestination vis-à-vis climate change, see Van den Brink and Van Urk, "Climate Change and God's Work of Election."

26. Berkhof, *Christian Faith*, 218; Muller, *Dictionary*, 46; by contrast, *articuli puri* (like, e.g., the doctrine of Trinity) can only be based on the specific sources of Christian faith and therefore lack such universal appeal.

27. Berkhof, *Christian Faith*, 217.

For others, however, this same belief has become one of the greatest stumbling blocks of the faith, as they find it impossible to square it with the manifold experiences of horrible suffering throughout history. In my view, Berkhof was right in stipulating that both views have to be qualified from a Christian perspective. Yet the dialectical relation between the two (one can easily imagine that they mutually elicit each other) compels us "to examine carefully the basis and the nature" of Christian belief in the providence of God.[28]

For early Christians, divine providence was not just clear from the (emerging) Bible, but it was forcefully confirmed by Platonic accounts of cosmic order and Stoic notions of divine *pronoia* (*pronoia* being the Greek equivalent of Latin *providentia*). Arguably, however, these philosophical influences came with a distortion of the authentic Christian way of confessing God's care. The concept of *pronoia* had strong deterministic and even fatalistic overtones: we can only patiently resign ourselves to the events that befall us and acquiesce to them. There is no place for notions of struggle, protest, or resistance, let alone intimations that human agency might decisively influence the course of things. In his very insightful monograph on the theology of providence, David Fergusson points out that Western theology in particular has been negatively affected by classical notions of *pronoia*, which are at odds with the pluriform ways in which Scripture portrays the God–world relationship: "Later Christian appropriations of ancient philosophy accommodated Stoic notions of patience, fortitude and perseverance, while also insisting upon the importance of viewing divine providence in its wider cosmic setting. But, in doing so, they did not fully reflect these scriptural themes and so set the theology of providence in a default position now in need of some revision."[29]

To illustrate this point, let us briefly explore the biblical passage that is usually seen as the fountainhead of all Jewish and Christian providence-talk, namely Genesis 22—one of the most unsettling and well-written narratives in the Bible.[30] Here, the Vulgate—the Latin Bible translation that was dominant in Europe for many centuries—renders the beginning of Abraham's answer to his son Isaac's question where the lamb was for the

---

28. Berkhof, *Christian Faith*, 217.

29. Fergusson, *The Providence of God*, 32. See also Wood, *Question of Providence*, 59: "Stoic themes and concepts were so much a part of the mental furniture of the ancient world that their assimilation into Christian thought and writing was largely unconscious." Therefore, Christians from the West should be open to criticisms and corrections by their fellow believers from other parts of the world whose worldviews have not been crafted in classical antiquity. What would a Christian view of providence that has *not* been co-shaped by Greek philosophy look like? I am grateful to Clive Pearson for this observation in response to an earlier draft of this chapter.

30. See Van der Kooi and Van den Brink, *Christian Dogmatics*, 234–35.

burned offering they were going to bring as follows: *Deus providebit* (v. 8). This literally means something like "the Lord will see to it." Indeed, the Hebrew word in the original text is the standard verb for "to see." Here it evidently has the meaning of "to provide" (thus most Bible translations, from the King James Version all the way down to the NRSVUE). What is interesting in this connection is that the notion of providence surfaces in what, despite the soberness of the narration, must have been experienced as a situation of utter anxiety and despair. Abraham had been summoned to offer Isaac, for whose wonderful birth he and his wife had so deeply longed. Could God really be so cruel and relentless as to reclaim this life through Abraham's own offering? Must Abraham become a child murderer?[31] But there seems to be no escape, for what else could Abraham do than to obey God? In that predicament, Abraham can only find a handhold in the faith that somehow, *God knows how*, a solution will emerge. It is only when there is nothing left that Abraham can do by himself that this appeal is made.

Obviously, it is a far stretch from confessing divine providence in such a situation of personal affliction and existential despair to turning it into a self-evident explanation of all worldly events. To be sure, the Bible contains other strands of providence literature, including texts that assume a universal scope of God's sovereign rule. Yet it is important to see how notions of providence emerged in the Jewish and Christian traditions in the first place. Like so many doctrinal notions, they were generalizations and, to some extent, abstractions of very concrete *experiences* that underlay them.[32] To be more precise: they arose in a context of existential threat and anxiety, and of an inexplicable trust that surfaced precisely in such conditions, somehow countering the tendency to give up and resign to utter desperation. In a sense, therefore, what is pertinent here is precisely the experience of *not* resigning and acquiescing but of finding courage to act in accordance with one's vocation, even in situations of extreme affliction. We may, of course, extrapolate from such experiences. For if trust in God is justified in such precarious conditions, it is certainly justified in all circumstances—which, in turn, suggests that "God holds the whole world in his hands." But even so, it makes a difference whether we endorse such a claim as an ossified theological truism or as an insight that encourages us in moments when we are on the brink of "freezing" in response to extremely challenging conditions.

---

31. One is reminded of Kierkegaard's reflections here, in his famous *Fear and Trembling* (1843).

32. I have previously argued this with regard to the (related) doctrine of divine almightiness. See Van den Brink, *Almighty God*, 176-78.

It should be clear that in the story of the binding of Isaac, we are at a far distance from Stoic notions of cosmic order and universal divine determinism. Even the faith that at the end of the day "God holds the whole world in his hands" does not imply that everything that happens in the world is willed by God so that all events and actions develop with iron necessity; it may rather go along with the acknowledgment that many events go *against* God's will. And indeed, the Bible is replete with stories about actions and events that flagrantly violate God's will. Yet this point has been largely obscured from view by the course taken by the Western theology of providence.[33] Fergusson points out that even though biblical interpretation led to some adjustment of Platonic and Stoic notions of providence—for example, the purposeful and fatherly will of God was not confused with blind fate—their tendency towards determinism and passivity became apparent and even dominant when the notion of providence was incorporated in the systematic unfolding of Christian doctrine in the medieval church in the West.[34] In this context, the concept of providence was closely tied to either the doctrine of God (or "theology proper"), where it came to be treated as on par with the eternal divine attributes, or to the doctrine of creation, where it was identified with God's ongoing preservation, concurrence, and governance of the world. Hence, it was isolated from further biblical themes like salvation, liberation, sanctification, and eschatological consummation.[35] In brief, we might say that belief in divine providence was never fully "Christianized" in the Western theological tradition.[36]

One way to address this problem is to become more attentive to what may be called the *particularities* of providence, that is, the variegated and multifaceted ways in which God deals with humans and with the world in the biblical literature. God is not just determining and governing what happens; God is also watching with grief how evil enters God's good creation; God is countering human sin, faced with opposition when doing

---

33. This theology is more encompassing than covering only the doctrine of providence. In post-Reformation scholasticism, subtle distinctions were elaborated between the hidden and revealed will of God, or the will of the (eternal) decree and the will of the (moral) precept, to make sure that all actions which in the biblical stories seem to be in conflict with God's will actually reflect that will at a deeper, hidden level. See Heppe, *Reformed Dogmatics* V, 25–27. Obviously, such distinctions were problematic insofar as they suggested a tension or even dualism within the divine will.

34. Fergusson, *Providence of God*, 58. Fergusson argues that there is a fundamental agreement between the positions of, e.g., Augustine, Aquinas, and Calvin and their followers on divine providence, so that we can speak of one "default Latin setting" that dominated the Western church (59–109).

35. Fergusson, *Providence of God*, 9–10. A clear and (in)famous example of this development can be found in the Heidelberg Catechism's treatment of divine providence in its Lord's Day 10, answer 27, where God's providence is equated with God's conservation and governing of "heaven and earth and all creatures."

36. See Van der Kooi and Van den Brink, *Christian Dogmatics*, 237.

so, therefore permitting many events which God does not want to take place, interacting and even negotiating with humans, responding to their prayers, liberating God's people from oppression, transforming them, abiding with them in their sufferings, promising them future consummation, and inspiring them to persevere in expecting the Kingdom that is arriving. Most of these patterns of God's providing action do *not* presuppose that "all is well," in the sense of completely unfolding according to God's will. In the biblical literature, particularly in its historical and narrative parts, there is a much more dynamic and often even dramatic interplay between divine and human actions and reactions. In order to grasp this, we have to bring into view the entire storyline of the gospel. As Conradie and Vaai rightly point out in their introduction to this volume, it is a danger "to separate God's care from, for example, God's work of salvation. That can only lead to theological speculation that becomes removed from the Christian gospel [...] and the Christological hinge around which the Christian faith turns."[37]

Obviously, however, it is difficult if not impossible to encapsulate the many particular ways of divine provision narrated in the Bible and succinctly mentioned above into a coherent system. That would almost inevitably mean to prioritize some of them (such as, traditionally, God's preserving, concurring, and directing activities) at the expense of others. We might therefore, as Fergusson suggests, rather prefer a "polyphonic" account of divine providence, which does not reduce the manifold patterns of divine involvement and care to only one or a few fixed models.[38] Yet in order to prevent polyphony from becoming cacophony, it is most helpful to rethink the doctrine of providence along *Trinitarian* lines, thus more fully "Christianizing" it. When confessing the providence of God, it is not the so-called "God of the philosophers" that Christians have in mind, that is, an inert, nonresponsive highest being or all-determining absolute principle, but the living God of Abraham who became incarnate in Jesus Christ in order to save the world and who indwells in humans and creation through his Spirit, transforming them and preparing them for the eschatological kingdom.[39] Following up on others who have elaborated a more fully

---

37. Conradie and Vaai, "In God We Trust?," 13.

38. For another recent attempt to construe a more pluriform account of divine providence in a strongly interdisciplinary context, see Silva and Kopf, *Divine and Human Providence*.

39. One may counter that elaborating the doctrine of providence in this direction comes at the cost of reducing its potential for interreligious dialogue and consensus-seeking. Would it not be great for Christians to have at least some core beliefs in common with Muslims (cf. their popular phrase *inshallah*), Hindus (cf. their *karma* principle), and people of other religious and spiritual orientations? To give in to this objection, however, would mean to resort to a notion of providence that is so vague and indiscriminate that it can be used and abused in all sorts of ways. See Webster, "On the Theology of Providence," 159: "[...] the task of a Christian theology of providence can only be undertaken by drawing upon the resources given to it by the gospel; it can only hear the prophets and apostles before it speaks—if not, it will have nothing to say."

Trinitarian account of providence, what follows are a couple of considerations sketching how this might be fleshed out.

First, even though it is tempting to couple the "traditional trio" to the three persons of the Trinity, it should no longer be taken for granted that divine providence can best be considered as comprising just the three aspects of preservation (*conservatio*), concurrence (*concursus*), and governance (*gubernatio*), as traditional scholastic treatises have it, and many contemporary accounts in their wake.[40] It seems that the selection of these three aspects in particular is largely based on the merger between Stoic and Christian influences mentioned above. That is not to deny that these three concepts represent some of the ways in which God does provide and care for us and for the world. For example, the notion of *conservatio* is of course of key importance given the precarious state of nonhuman nature in the "Anthropocene": if God is active in preserving nonhuman nature, then perhaps we should be so as well. But these three perspectives are not the only ones from which to view divine providence, and there is actually no theological reason to single them out for special attention. The fact that all three of them in some way back up the status quo—God sustaining both the nonhuman creation and human agency and watching over them—compels us to add to them equally relevant concepts that are more forward-looking and therefore more in line with the eschatological orientation of the gospel. Thus, rather than limiting ourselves to the traditional trio in explaining how God provides, we might consider complementing them with concepts like interruption, liberation, and transformation (to be briefly elaborated below).

Second, the doctrine of providence should not be exclusively related to either the treatise *de Deo uno* (i.e., God's being conceived of "before" any Trinitarian differentiation) or the doctrine of creation, as if it were just an appendix to the first article of the creed. Instead, providence "covers the full range of Christian doctrines" and is "variously appropriated to Father, Son and Spirit."[41] Fergusson even creatively suggests that providence should not be treated as a distinct *locus* in and of itself, but "as an element shared and modulated by all the doctrines," even to the extent that it is "parasitic" on them.[42] If we want to have a dogmatic label that comprises

---

40. Examples of contemporary theologians who stick to this threefold scheme (indeed often ascribing one of the three to each of the Trinitarian persons) include Migliore, *Faith Seeking Understanding*, 125–27; Wood, *Question of Providence*, 78–91; Higton, *Christian Doctrine*, 190–211; Van der Kooi and Van den Brink, *Christian Dogmatics*, 239–44 (where a Trinitarian interpretation of each of the three is suggested); Conradie, "Common Grace and Sustainability" (in this volume). It is the genius of Fergusson's account to question the appropriateness of fixating our view on these three aspects of the much more polyphonic biblical witness.

41. Fergusson, *Providence of God*, 297.

42. Fergusson, *Providence of God*, 297.

God's actions which extend the goodness of creation over time, the concept of "common grace" might be more appropriate. Providence is a much more encompassing notion, which should not be relegated to either the doctrine of God or the doctrine of creation, as in that case "it suffers from constraints that inevitably distort its expression within other doctrinal loci."[43] Indeed, it will be hard to take seriously from a theological point of view the various ways God's loving care functions in these other loci when providence has already been "stored" in either of these doctrinal "boxes".

Third, it follows that belief in divine providence has to be tied closely to and even follows from (rather than preceding) belief in God the Father of Jesus Christ, who came into the world to serve rather than to be served (Mark 10:45). It is the God who identifies with this Jewish man Jesus in such a radical way as exemplified by the hypostatic union, whose providence Christians confess. From the outset, it is therefore out of the question that the historical appeals to divine providence to sanction the emergence of the Third Reich could be legitimate—such appeals can only be rejected as idolatrous distortions of sound Christian doctrine. They also serve as a warning of how easily belief in divine providence, when isolated from its biblical context, can be turned into brutal ideologies that in fact run counter to the gospel. Even after the Second World War, the world has seen many instances of such distorting and false political appeals to providence—and continues to see them right now (these lines are written while the Russian war against Ukraine is raging—a war that is justified by strongly theological accounts and especially by appeals to divine providence).

Fourth, when focusing on the providing work of God the Father, we may consider *interruption* as a key term.[44] Whoever peruses the biblical narrative will observe that time and again, God is interrupting the normal course of events. Whereas people usually tend to move on in paths and habits they have become used to, God comes down to confuse their plans and thoughts (Gen 11). In fact, throughout the Old Testament, God continues to send God's judgments and prophets in order to prompt people to change their habitual ways of life. Sometimes they do (as testified in the wisdom literature), but mostly they refuse to do so. The New Testament picks up on this by witnessing that "in the fullness of time" (Gal 4:4), God finally made a decisive interruption in the normal course of history by sending God's Son.

---

43. Fergusson, *Providence of God*, 297.

44. The notion of interruption has in particular been highlighted as an important theological category by Eberhard Jüngel. See, e.g., his "The Truth of Life"; David Nelson, *The Interruptive Word*. Note that "interruption" is different from the "disruption" that is also sometimes ascribed to God (cf. Ernst Conradie's essay in this volume). Interruption is not in itself destructive, as it leaves room for human freedom to act upon it. It is only when, time and again, divine interruptions have been left unheeded that total disruption threatens.

God provides by interrupting and calling us back from wrong patterns of behavior. The climax of this reading of the story of providence can be found in the parable of the vineyard (or the wicked tenants; Matt 21:33–45), where the owner interrupts the state of affairs again and again by sending his servants, in order to finally send his own son.

Fifth, this leads us quite naturally to consider the *liberating* way in which the incarnate Son provides on our behalf.[45] To call this way liberating is of course far from saying anything new or provocative (apart from the provocative character of the gospel itself). The church was born from the very confession of Jesus as Lord and Savior—that is, Liberator. In the gospel narrative, the manifold actions in which Jesus sets people free, psychically, physically, and spiritually, find an unexpected ending in his crucifixion. Yet as the drama unfolds, it turns out that it is through Jesus's self-offering at the cross that God provides for our reconciliation with God, with each other, and with our own past in a way that we could never have imagined and can still adore more than we can fathom it. Christians experience themselves as having been freed by Jesus' liberating work to become members of a new community that lives out this freedom and invites others to let themselves be liberated as well. At the same time, Jesus' liberating work is seen as fully in line with Old Testament talk about God. For example, the story of Isaac's binding, with which we started, can also be read as the story of Isaac's liberation, typologically prefiguring the liberation brought about by Christ. When Christ is called the Lamb of God, this probably refers to the lamb provided by God to Abraham as the burnt offering he was about to bring in Isaac's place (Gen 22:8).

Sixth, the New Testament depicts the role of the Holy Spirit as utterly *transformative* and forward-looking. To be sure, the Spirit provides comfort and support—but not to acquiesce in some unjust *status quo*; rather, it is in the specific situation in which Jesus' followers have to move on without direct access to Jesus that this comfort and support is provided (John 14–16). The main role of the Spirit according to the New Testament is much more challenging and forward-pushing, as, for example, the Dutch theologian Oepke Noordmans has highlighted.[46] In pushing the gospel beyond all sorts of human boundaries throughout the world (see especially the Book of Acts), the Spirit transforms people (and places) by redirecting their lives towards the Kingdom of God. The Spirit provides for us not by leaving us as we are and where we are but by purifying and renewing us, loosening the ties with which we are bound to the mundane, money-driven world.

---

45. For a deliberate attempt to link traditional talk of divine providence to "liberative theologies," explicating divine struggle as "a feature of providence," see Hayes, "Discerning Providence."

46. Unfortunately, hardly any work of Noordmans has been translated into English; but cf. Blei, *Oepke Noordmans*, especially chapter 15.

It is important to conceive of divine providence as incorporating this change-oriented pneumatological perspective, which does not only affect our individual lives but also creation at large. Indeed, nowhere is the forward-pushing character of the work of the Spirit so palpable as in Romans 8, where the fate of the believers—the ones who bear "the first fruits of the Spirit" (8:23)—is seen as inextricably connected with that of a nonhuman creation that is groaning in labor pains. Here, God's loving care and providence even take the form of helping us in our weakness by empowering us to share the burdens of a biosphere that is suffering from its bondage to decay until its future consummation (see also Rev 22:17).

Finally, re-envisioning providence along such lines that are taken from the overall biblical narrative—in fact, such lines can be elucidated much more clearly by retelling the relevant stories than by introducing a new set of theological concepts—may also mean that we have to revisit the classical concepts of conservation, concurrence, and governance so as to orient these as well to the plot of salvation history as narrated in the Bible. For example, we may follow David Fergusson in more robustly maintaining the distinction between the divine will and divine permission than has been done in the Western tradition. God is not indiscriminately at work always and everywhere but in very particular ways and places—and the gospel sharpens our eyes to discern more perceptively where and how.[47] In line with this, the governance of God might be constructed "in promissory terms, rather than in a total control that is everywhere and always exercised."[48] Similar readjustments could be envisaged with regard to the concepts of preservation and concurrence, both of which need not necessarily be seen as sanctioning a status quo but can be conceived of as directed to an eschatological future that is engendered by God's interrupting, liberating, and transformative providence.

When re-envisioning God's providence and care for us along such lines, it is important to emphasize that this triune, trustworthy God can only be known through faith.[49] This is especially important when we wonder how God's providence and loving care can be reconciled with the suffering of God's creatures. The truth is that there is no rational or theoretical way in which this can be done.[50] Theodicies, of whatever nature they are, and however helpful they may be in finding "fragments of meaning" amid suffering, will not be able to offer full satisfaction. The divine providence

---

47. Brümmer, *Speaking of a Personal God*, 125–27.

48. Fergusson, *The Providence of God*, 300.

49. See Pearson, "God's Continued Providence," 399: "Its strength and resilience has lain in its being a confession of faith and, as such, a revelatory claim."

50. See Elizabeth Pyne's highlighting of this point in her contribution to this volume.

cannot be vindicated or demonstrated by empirical observation, but it has its unique setting in the trust that has been awakened by the God whom we have come to know in the life, death, and resurrection of Jesus Christ and who transforms us through his Spirit in the community of the faithful. Apart from this deeply personal and existential—but also communal—relationship of trust, the specifically Christian view of providence cannot come off the ground, and we will easily slip back into flat sub-Christian notions (such as "nothing happens by accident," "one should accept one's fate," "we are our brain," or "life is entirely meaningless") that happen to be culturally dominant. Arguably, it is such *secularized* notions of providence that form the real danger, given the spiritual transformation of humans that the "Anthropocene" requires.

■ The Prospects of Providence in the "Anthropocene"

Returning now to the setting of the "Anthropocene" with which we started, and which forms the background of this volume and series, what does our brief inquiry into the particulars of providence mean for the ways we may and may not appeal to God's care in a time of global ecological disaster? As we have seen above, an appeal to divine providence is often made as an attempt to suggest that we do not have to change our patterns of behavior in response to such threats, since "God will provide." In particular, when it comes to the issue of climate change, such an appeal is sometimes readily made because the climate is often seen as a divine domain par excellence.[51] Did not God indeed make the promise which we encountered above to Noah and his family: "As long as the earth endures, seedtime and harvest, cold and heat, summer and winter, day and night shall not cease" (Gen 8:22, NRSVUE)?[52] And are God's covenant promises not irrevocable? Indeed, the promise made here is unilateral and unconditional. Since the Lord had concluded that "the inclination of the human heart is evil from youth" (8:21; cf. 6:5), the Lord's promise is deliberately not made dependent on human cooperation.

Yet even so, this does not imply that in our contemporary situation we need not bother about climate change and its consequences. As we have seen, as long as we can make a difference by taking appropriate action, God's providence can never be appealed to for resigning against unjust situations. The world's changing climate evidently goes hand in hand with an unfair distribution of its consequences. Nor can the appeal to God's

---

51. Van Urk, "De samenhang bij christenen," 112–13.

52. The remainder of this paragraph is adapted from my essay "The Role of Scripture and Theology."

providence be used as a license for human brinkmanship, as if we could take risks by being sluggish in addressing the situation because of our trust in God. Usually, we take this point for granted. For example, when reading in Psalm 121, "The Lord will keep you from all evil. The Lord will keep your going out and your coming in" (121:8, NRSVUE), nobody will conclude from such words that we do not need to use seatbelts in our cars—let alone that it is a sign of unbelief or ingratitude to do so. On the contrary, we realize that such words do not release us from the obligation to take care of ourselves. Thus, it would be equally strange to conclude from Genesis 8 that taking climate change seriously is a lack of trusting and acknowledging God, as some have claimed.[53] Instead, we might better argue that if a stable climate matters so much to God, it definitely should be *our* concern as well.[54]

In this connection, it is helpful to focus on the aspects of divine providence we distilled from the biblical storyline in the previous section. If God's care for us and for creation at large can be *interrupting*, the threatening global climate crisis may be a signal of this—an "Old Testament" type of prophetic warning and judgment that we have to change our ways; otherwise, we will lose not only our relationship with God but also our land. If God's care for us and for creation at large is *liberating*, this is definitely, in the first place, good news for those who are the primary victims of today's climate change, humans and nonhumans alike (for yes—God *does* also care for the oxen). Rather than resigning to the status quo, Christians should stand with them and identify themselves with their just case. And finally, if God's care for us and creation at large is *transformative*, we are being pushed forward by the divine providence, out of the comfort zone of a consumerist lifestyle which is typical for the old world but which will have no place in the new world that is to come. We are even prompted by the Spirit to feel the pains of ecological destruction as our own pains rather than isolating ourselves from nonhuman creation (Rom 8). Calvin already argued that the most special form of God's providence is the one by which "He governs His faithful ones, living and reigning in them by His Holy Spirit."[55]

---

53. Cornwall Alliance, "Renewed Call to Truth," 17 (see above, notes 19 and 20).

54. Here, the scholastic notion of *concursus* retains its relevance, in that it does not allow us to see human and divine action as mutually outdoing each other. I owe this observation to my colleague Henk van den Belt. See also Kathryn Tanner's "modern classic," *God and Creation in Christian Theology*.

55. Calvin, *Against the Libertines* (1545), 191. Calvin makes a tripartite distinction here (and only here, by the way) between a universal dimension of divine providence that encompasses the natural order, a special aspect that is focused on human history, and a "third species" that concerns the faithful. See Reardon, "Calvin on Providence," 529–33: "This third kind of Providence Calvin identifies with the presence of the Holy Spirit in the People of God" (532).

As said, all this does not mean that there is no legitimate place for notions of divine conservation, concursus, and governance. It is more appropriate, however, to remind ourselves of these aspects of divine providence *after* the other ones instead of starting with them. For these classical aspects teach us that when we humans have taken God's interruptive work seriously (e.g., by actively responding to the ever more indisputable and grim signals of anthropogenic climate change), when we have become involved in mirroring Christ's liberating work to the victims of this world's power dynamics, and when we "let the Lord work in me through his Spirit"[56] so as to become part of his transformative work—*then* we may trust that all this is not in vain, since it is not us but God who will steer the world towards its eschatological future.

To paraphrase Bonhoeffer's famous saying that "only those who cry out for the Jews may sing Gregorian chants," we might consider that only those who participate in God's interruptive, liberating, and transformative action can put their trust in God's conservational, concurring, and governing guidance of the world and its climate.[57] This also means that Christians who take divine providence in all its facets seriously need not succumb to so-called climate despair, the utter desperation of some secular climate activists (and the "freezing" of others) who fear that the world's climate system is now destined to a total collapse. For Christians, however dire the prospects may become, there remains hope.[58] This hope is grounded in the story about the one who brought about new embodied life through the utter desperation of his death on the cross. The rumor goes that this new life will eventually not be anthropocentric but encompass all subjects of God's care and concern, including nonhuman forms of life and the whole entangled web of a renewed creation.

■ Bibliography

Bauckham, Richard. *Living with Other Creatures. Green Exegesis and Theology*. Waco: Baylor University Press, 2011.

Berkhof, Hendrikus. *Christian Faith: An Introduction to the Study of the Faith*. Revised edition. Grand Rapids: Eerdmans, 1986.

Bethge, Eberhard. *Dietrich Bonhoeffer: Man of Vision, Man of Courage*. New York: Harper & Row, 1970.

Blei, Karel. *Oepke Noordmans: Theologian of the Holy Spirit*. Grand Rapids: Eerdmans, 2013.

---

56. Heidelberg Catechism, Lord's Day 38, answer 103.

57. Bonhoeffer's dictum cannot be found in his work but goes back to reports of his students. See Bethge, *Dietrich Bonhoeffer*, 512.

58. On the Judeo-Christian concept of hope and its significance in times of climate uncertainty, see Hasselaar, *Climate Change, Radical Uncertainty and Hope*, 65–92 and Van den Heuvel, "Hope."

Blunt, Sheryll Henderson. "Cool on Climate Change." *Christianity Today* 50:10 (2006) 26–27. http://www.christianitytoday.com/ct/2006/october/8.26.html.

Brümmer, Vincent. *Speaking of a Personal God: An Essay in Philosophical Theology*. Cambridge: Cambridge University Press, 1992.

Calvin, John. "Treatise against the Libertines (1545)." In *Ioannis Calvini Opera quae supersunt omnia*, edited by Johann-Wilhelm Baum, Edouard Cunitz, and Eduard W.E. Reuss, Vol. 7, 145–248. Brunswick: Schletschke, 1863–1900.

Cameron, Helen, Deborah Bhatti, Catherine Duce, James Sweeney, and Clare Watkins. *Talking About God in Practice: Theological Action Research and Practical Theology*. London: SCM, 2010.

Clough, David. "The Bible and Animal Theology." In *The Oxford Handbook to the Bible and Ecology*, edited by Hilary Marlow and Mark Harris, 401–12. Oxford: Oxford University Press, 2022.

Cornwall Alliance. *A Renewed Call to Truth, Prudence, and Protection of the Poor: An Evangelical Examination of the Theology, Science, and Economics of Global Warming*. Cornwall Alliance: For the Stewardship of Creation, 2009. https://www.cornwallalliance.org/docs/a-renewed-call-to-truth-prudence-and-protection-of-the-poor.pdf.

Deane-Drummond, Celia E. *A Primer in Ecotheology: Theology for a Fragile Earth*. Eugene: Cascade, 2017.

Fergusson, David. *The Providence of God: A Polyphonic Approach*. Cambridge: Cambridge University Press, 2018.

Gilmour, Michael J. *Eden's Other Residents. The Bible and Animals*. Eugene: Cascade, 2014.

Grudem, Wayne. *Politics—According to the Bible: A Comprehensive Resource for Understanding Modern Political Issues in Light of Scripture*. Grand Rapids: Zondervan, 2010.

Ham, Ken, and Bodie Hodge, eds. *The New Answers Book 4*. Green Forest: Master Books, 2013.

Hasselaar, Jan Jorrit. *Climate Change, Radical Uncertainty, and Hope: Theology and Economics in Conversation*. Amsterdam: Amsterdam University Press, 2022.

Hayes, Andrew. "Discerning Providence: How the Reign of God in Liberation Theology Explicates Divine Struggle as a Feature of Providence." *Modern Theology* 39 (2023) 435–54.

Heppe, Heinrich. *Reformed Dogmatics Set Out and Illustrated from the Sources*, Revised and edited by Ernst Bizer. Grand Rapids: Baker, 1978.

Higton, Mike. *Christian Doctrine*. SCM Core Text. London: SCM, 2008.

Horrell, David G., et al., eds. *Ecological Hermeneutics: Biblical, Historical and Theological Perspectives*. London: T&T Clark, 2010.

Jüngel, Eberhard. "The Truth of Life: Observations on Truth as the Interruption of the Continuity of Life." In *Creation, Christ, and Culture: Studies in Honour of T.F. Torrance*, edited by R. W. A. McKinney, 231–36. Edinburgh: T&T Clark, 1976.

Kierkegaard, Sören. *Fear and Trembling. Dialectic Lyric by Johannes de Silentio*. London: Penguin, 1985 (1843).

Linzey, Andrew. *Christianity and the Rights of Animals*. Eugene: Wipf & Stock, 2016.

Lowe, Benjamin S., Rachel L. Lamb, and Noah J. Toly. "Climate Skepticism, Politics, and the Bible." In *The Oxford Handbook of the Bible and Ecology*, edited by Hilary Marlow and Mark Harris, 425–44. Oxford: Oxford University Press, 2022.

Migliore, Daniel L. *Faith Seeking Understanding. An Introduction to Christian Theology*. 2nd ed. Grand Rapids: Eerdmans, 2004.

Milman, Oliver. "Climate Denial Is Waning on the Right: What's Replacing It Might Be Just as Scary." *The Guardian*, November 21, 2021. https://www.theguardian.com/environment/2021/nov/21/climate-denial-far-right-immigration.

Muller, Richard A. *Dictionary of Latin and Greek Theological Terms*. Grand Rapids, MI: Baker, 1985.

Nelson, David. *The Interruptive Word: Eberhard Jüngel on the Sacramental Structure of God's Relation to the World*. London: Bloomsbury, 2014.

Pearson, Clive. "God's Continued Providence." In *Christian Theology and Climate Change*, edited by Ernst M. Conradie and Hilda P. Koster, 395–405. London: T&T Clark, 2020.

Preece, Rod, and David Fraser. "The Status of Animals in Biblical and Christian Thought: A Study in Colliding Values." *Society & Animals* 8:3 (2000) 245–63.

Reardon, P. H. "Calvin on Providence: The Development of an Insight." *Scottish Journal of Theology* 28 (1975) 517–34.

Schweitzer, Albert. *The Mysticism of Paul the Apostle*. New York: Seabury, 1968.

Sideris, Lisa H. "A Rupture of the Earth. An Implicit Augustinian Theology of the Anthropocene." In *Theology on a Defiant Earth: Seeking Hope in the Anthropocene*, edited by Peter Walker and Jonathan Cole, 17–39. Lanham: Lexington Books, 2022.

Silva, Ignacio, and Simon Maria Kopf, eds. *Divine and Human Providence: Philosophical, Psychological and Theological Perspectives*. London: Routledge, 2021.

Singer, Peter. *Animal Liberation: A New Ethics for Our Treatment of Animals*. New York: New York Review Book, 1975.

Tanner, Kathryn. *God and Creation in Christian Theology: Tyranny or Empowerment?* Oxford: Blackwell, 1988.

Van den Brink, Gijsbert. *Almighty God: A Study of the Doctrine of Divine Omnipotence*. Kampen: Kok Pharos, 1993.

———. *Reformed Theology and Evolutionary Theory*. Grand Rapids: Eerdmans, 2020.

———. "The Role of Scripture and Theology in Addressing Climate Science Skepticism", in: Michael Borowski, Tomas Bokedal & Ludger Jansen (eds.), *Science, Scripture & Theology: Moving Beyond the Impasses* (Tübingen: Mohr Siebeck, 2025 forthcoming).

Van den Brink, Gijsbert and Eva van Urk. "Climate Change and God's Work of Election." In *Christian Theology and Climate Change*, edited by Ernst M. Conradie and Hilda P. Koster, 451–61. London: T&T Clark, 2020.

Van den Heuvel, Steven. "Hope: An Essential Capacity for Human Development." In *Driven by Hope: Economics and Theology in Dialogue*, edited by Steven van den Heuvel and Patrick Nullens, 199-210. Leuven: Peeters, 2018.

Van der Kooi, Cornelis, and Gijsbert van den Brink. *Christian Dogmatics: An Introduction*. Grand Rapids: Eerdmans, 2017.

Van Urk-Coster, Eva. "De samenhang bij christenen tussen evolutie- en klimaatstandpunten [The Connection between Views on Evolution and Climate Change among Christians]." *Radix* 46:2 (2020) 108–18.

———, ed. "Our Current Extinction Crisis: Ethical and Theological Perspectives." *Journal of Reformed Theology* 17:2 (2023) 121–215.

Veldman, Robin Globus. *The Gospel of Climate Skepticism: Why Evangelical Christians Oppose Action on Climate Science*. Berkeley: University of California Press, 2019.

Webster, John. "On the Theology of Providence." In *The Providence of God: Deus Habet Consilium*, edited by Francesca Aran Murphy and Philip G. Ziegler, 158–75. London: T&T Clark, 2009.

White, Alan. "Climate Change Facts: Should We Be Concerned?" In *The New Answers Book 4. Over 30 Questions on Evolution/Creation and the Bible*, edited by Ken Ham, 187–98. Green Forest: Master Books, 2013. https://answersingenesis.org/environmental-science/climate-change/should-we-be-concerned-about-climate-change/.

Wirzba, Norman. *This Sacred Life: Humanity's Place in A Wounded World*. Cambridge: Cambridge University Press, 2021.

Wood, Charles M. *The Question of Providence*. Louisville: WJK, 2008.

Continuing the Conversation in Christian Ecotheology on God's Providence

EMC and ULV: The series on *An Earthed Faith* draws on various strands of narrative theology from around the world. The assumption is that the Christian faith has a narrative shape and structure. It tells a story—interpreted differently in different contexts—of who God is and what God has done, is doing, and may do within our world. Do you have any comments in this regard given the various contributions to this volume?

GvdB: I very much appreciate the rich variety of perspectives included in this volume, covering different continents, denominations, genders, ages—and I am sure also political and socio-economic conditions. If it is true that it is only "along with all the saints" that we can fathom the various dimensions of the love of Christ (Eph 3:18–19), that same principle will also apply to the providential care of the Father. So we really need each other's stories and reflections to get a glimpse of the multifarious ways in which God can be seen to be acting in the world. In fact, one can even wonder whether the notion of a unilinear story (in the singular?) can capture this wonderful plurality. Metaphors like those of a polyphonic musical composition or a richly polychromatic tapestry (both of which exhibit clear

How to cite: Conradie, EM & Vaai, UL et al. 2024, 'Continuing the Conversation in Christian Ecotheology on God's Providence', in EM Conradie & UL Vaai (eds.), *Making Room for the Story to Continue?*, An Earthed Faith: Telling the Story amid the "Anthropocene", vol. 4, AOSIS Books, Cape Town, pp. 235–254. https://doi.org/10.4102/aosis.2024.BK415.11

unifying patterns) may help here. Anyhow, I am grateful to the editors for gathering such a varied chorus of different voices that continue to tell the stories of God's dealings with the world in the context of the global predicament of the "Anthropocene."

EMC: Yes, Gijsbert, these are stories in the plural. However, the opposite danger is that such stories become disconnected from each other. It is not only about losing touch with one another ecumenically. The danger is also that God's work of salvation becomes disconnected from God's work of creation—so that salvation is understood as salvation from creation!

ULV: Yes, thanks Gijsbert. I think the pluralistic element of stories that reflect the multifarious ways of God's work in the world is a wonderful point to consider. What I fear is this: on the one hand, imperial rule has always been about the establishment of a world under the rule of the "Onefication" political and economic system, the belief that there is only one constituent source of goodness. The elements of this Onefication narrative are also reflected in many religions, in particular Christianity. On the other hand, I also fear, likewise, a pluralistic society that seems to stretch God's salvation to the point that it becomes too complex, where there is no unifying and interweaving reality.

ASA: I agree with Gijsbert that listening to various voices in this volume gives us a glimpse of the diverse and universal unity of the Church. It is also true that such various contexts tend to polarize the body of Christ. However, while glancing through other essays, I see the universality of God's providence over creation and how history has been understood to be shaped by God. Notwithstanding ecological woes in the "Anthropocene," it seems we all still hold the notion of divine providence.

CP: The hearing of different narratives is intriguing in the "Anthropocene": how does the contextual intersect with the planetary? I suspect that we are inclined to privilege the local (which is understandable) as the planetary invites us to be interdisciplinary. Seems like a much larger "superproblem" and may necessitate a greater engagement with other disciplines.

NHC: I think it is important to emphasize that the "Anthropocene" is a new problem because of its scope, but it is also not new in that we have lived through a multitude of tragedies and horrendous evils in our history and, as well, in the last few thousand years. It is important to hold onto the idea that we are not alone in our work for the healing of the planet and adaptation to a changing climate; the creator Spirit has always been with us and in nature's development. The Spirit may surprise us yet.

ER: Yes, Nicola, this collection of stories about the work of God from around the world testifies that God is working wherever the "Anthropocene" affects God's people, without any boundaries. I pray that these narratives continue to find room around the world in order to affect the change that is needed in behavior, lifestyle, and attitudes for the sake of preserving earth as our "common home."

EMC and ULV: For this volume, the focus is on God's work of providence. The question that we addressed was this: "How could the suffering of God's creatures in the 'Anthropocene' be reconciled with trust in God's loving care?" Would you agree that the rupture of the "Anthropocene" radicalizes the challenge to any Christian affirmation of God's care? Why?

ER: Yes, the suffering of God's creatures calls for the salvation of all creatures. Such salvation has required the suffering of Godself (in Jesus Christ) on the cross. Human beings are complicit in such suffering, as implied in the term "Anthropocene."

GvdB: I find this a difficult question to address. For haven't we seen many horrors in the recent and not-so-recent past which provoked the very same question? Yet if belief in God's care survived the Holocaust and other genocides and atrocities, we may well trust that it will survive anything, including the "Anthropocene" (provided that *we* survive it). Skeptics would say (and the logical positivists of the 1960s actually did say) that this is precisely the problem: no matter what happens, nothing is taken by believers as a falsification of their trust in God. If we look into the "cradle" of belief in providence, though, this is not so strange, since it precisely emerged and started to function in situations of utter despair. When everything falls apart, hope in God was the only thing that kept the faithful going. One might just as well say, therefore, that the "Anthropocene" rekindles the hope for divine providence and care instead of extinguishing it.

EMP: Having been part of a number of conversations and projects that focus on the "Anthropocene" as a new context for theology and theory, and having often endeavored to persuade others of the significance of the changes it has wrought, I was somewhat surprised in writing this essay to find that I struggled to pinpoint any satisfactory answer regarding what, finally, is different about the "Anthropocene" specifically with respect to creaturely suffering and the reality of God's providence. I thus want to echo aspects of what Gijsbert says about the continuity of this challenge with previous events or periods, yet with a twist. Instead of focusing on the faith that persists "on the other side," as it were, of these atrocities and catastrophes, is there something to be gained from reflecting on them—from dwelling in them, maybe—precisely as ruptures or radicalizing moments? To be more direct: I am worried about the potential for claims about the unprecedented scope and scale of the "Anthropocene" to displace histories of oppression, victimization, and loss—including loss of possible earth-futures, the destruction of worlds. Now, this does not *necessarily* follow, and there may be ways to counteract the risk, including prioritizing diversity among the voices that tell and critique and retell the Christian story (this volume may be counted as one effort); tying scientific conceptualizations of the "Anthropocene" to humanities-led ones; and elaborating the connections (which is not to say absolute correspondence) of this global epoch to systems such as European colonization, capitalism, and fascism.

ASA: As someone who has been wrestling with the problem of evil, I find the main focus of this volume raising the same difficult questions as those asked by critics of theism in general and Christianity in particular with respect to the problem of evil, given the Christian concept of God. Further, ecological woes in the "Anthropocene" seem to suggest that God does not exercise meticulous divine providence over the creation. Someone who finds the belief in God ludicrous could argue that given the rupturing of the ecosystem in the "Anthropocene," the great-making properties of God, love, and care for the creation as upheld in theism raise many questions and objections. However, a theist may want to argue that just as the presence of evil in general does not disprove the existence of a loving and caring God, so does the issue of ecological destruction. In other words, the ecological woes in the "Anthropocene" do not cast doubt on God's love and care for creation. Instead, they provide support of that because, if not for God's love and care for creation, the world would have ended long ago.

CP: I think the "Anthropocene" (even though it is not yet accepted by geologists) is best understood as a rupture. Here, I am thinking of it in terms of the emergence of the Earth System sciences: the rupture is via changes to the system—to the climate, the emergence of extreme weather beyond what we have had to cope with in the past. It is a rupture, a *kairos* moment; I think our theological traditions unfolded in a very different earth/climate system. It is a rupture for low-lying islands that will become uninhabitable or be lost to rising sea levels. With many other traumas, the land, the weather / the atmosphere upon which life depends remained. That is not to say that historical injustices are not real or should be denied. Some writers now refer to the sheer complexity of climate justice because of changes to the Earth System and the legacy and current reality of historical grievance.

NHC: Yes, Clive, but one also needs to consider issues of biodiversity in the "Anthropocene." A deep knowledge of both biology and physics are incentives to reformulate and revision both our theology and our biology. Theology must be done with the full ambiguity of diverse hominid species and the deep continuities between human and other animals in mind. Biology must acknowledge that there is no longer a consensus for a blind materialism. To go down this route is to be convinced of the love of God for all species and all life, and of our responsibility to care for that life. None of this solves the problem of evil in itself, but it does place theodicy in a context that is more holistic and more expansive than was previously the case.

GvdB: I'd like to second Nicola here—these seem to me very pertinent insights in the contemporary scope and tasks of theology and the sciences. I would complement (in line with Clive) that perhaps we should add climate science, and/or Earth System science, to the list of traditional sciences (physics, biology, etc.) with which a realistic theology has to engage.

MS: I agree with Gijsbert that human beings, like other species, have endured a vast number of atrocities and sufferings and that the difficulties of the Anthropocene will be a new circumstance that will lead us as believers to trust in God's providence. On the other hand, I have difficulty quantifying and, more importantly, qualifying suffering. The suffering of those who suffer is immeasurable (and, as Nicola observes, this is about all species and all life). Undoubtedly, the complexity of the events that fall under the umbrella of this age, as well as the various dimensions in which the related sufferings present themselves, is a separate phenomenon that has resulted in the development of several academic and scientific approaches. Perhaps in the future, we will be able to identify a clear-cut new paradigm of knowledge—and also of theology. However, I believe that the Christian affirmation of God's care is based on core aspects of the Christian faith that do not change; it is up to us to creatively narrate them each time from our distinct contexts.

ULV: I do feel that while the conversation and the volume head more into God's providence and God's care, and the question on how we can trace that divine care in the suffering of the people, which is important to articulate amid the risks of the Anthropocene, more should be focused on the theologies of human transformative responsibilities, ethics of care, and philosophies of restraint.

EMC and ULV: Would you agree that there is a tendency in contemporary Christian theology to shy away from the theme of providence, even though trust in God's ongoing care is core to the daily praxis of Christians, in particular Indigenous communities? Why does the language of providence sound so unfamiliar outside Western traditions? Why are Christians often more obsessed with God's punishment than with God's care?

MS: In my experience, I find that the theme of providence is avoided in a number of ecclesial and academic theological communities. My essay focuses on some aspects of this phenomenon. I disagree with the assertion that the language of providence is unfamiliar outside of Western traditions. In the Indigenous traditions of South America—my immediate context—

there are fundamental beliefs about the care and provision for life by the divine. In turn, the role of human beings in the recognition and care of these gifts is well defined. On the other hand, in the context of secularized Western traditions, I find that the language of providence is often set aside because it appears naïve and/or irresponsible. As for the question of punishment versus God's care, the emphasis on God's punishment still persists in some conservative or fundamentalist communities. Yet what I observe, mostly in my immediate context, is that the issue of suffering linked to the "Anthropocene" appears as related to actions that have consequences and not to the issue of "crime and punishment."

GvdB: As to the first question, yes, perhaps that's true—even though it seems that contemporary theology is experiencing a kind of resurgence of interest in the theme of providence. Theology's tendency to shy away from the topic may have to do with the fact that, post-Hitler and post-apartheid, theologians have become aware of the many dangerous ideological abuses of providence-talk. In the daily praxis of many Christians—not only in Indigenous communities, I would say, but even among "ordinary" Western believers—talk of God's care and providence, on the other hand, is still prominent (e.g., evangelical Christians telling many authentic stories about the wonderful ways in which God cared for them). Even in secularized forms, providence-talk is often easy to find, albeit in a fairly generalized way (such as in sentences like "nothing happens by accident" and "someone high-in-the-sky must be watching over me"). Awareness of "the pitfalls of providence" seems less developed among them. I don't know whether the observations that Christians are more obsessed with God's punishment than with God's care also apply to the West. More and more Western Christians rather believe that God hardly punishes anymore and that hell may be empty (without realizing that the "Anthropocene" may bring hell back on the scene …). I leave it to others to tell whether or not this may be a difference between Western and other Christians (e.g., among Indigenous people).

ASA: As an African from an Islamic-dominated region of Nigeria, the language of providence is quite familiar to me. In fact, all of life is viewed as working ultimately due to the divine finger. As for the question of punishment rather than care, in my view, the majority of Christians in Africa are not necessarily legalistic, therefore seeing an avenging or angry God who is obsessed with punishing sinners. But I think the main issue lies in the application of Deuteronomistic theology (which resonates with the African worldview of cause and effect) to every situation, to the extent that more attention is now given to the curses emanating from disobedience, with less emphasis on God's love and care towards us. To add to that, given the emphasis on dramatic experiences or the miraculous, it sometimes appears as if God is not caring for creation when nothing dramatic happens as people may have anticipated.

ULV: There is a kind of uncomfortable, one-dimensional, straightforward providence led by some church traditions where the question of God's providence is no longer complex and multidimensional, driven by mere rationalization. The complexity of the issue must not be overlooked. In Indigenous communities, God's providence is deeply disconnected from the everyday life simply due to its academia orientation, and most of the conversation around providence is driven by either fundamentalist movements that hold a prominent presence or churches that are not ready to review their theological ideas. But perhaps the most challenging [part] of the providence story is that its articulation is mainly influenced by the Western philosophical tradition that continues to shape Christian doctrines such as this one.

ER: God's providence is not the prerogative of Western people because God's work, care, and love have no limits. God has been revealed to people of all nations in different ways. It has even reached Rwanda, as Rwandan names such as Hatangimana (God provides for) and Harerimana (God takes care of our children) indicate. God's providence is experienced everywhere around the world, as it has been testified through the stories collected in this volume.

CP: Yes, Eraste, but I need to admit that the theme of providence has been seldom referred to in my context—though I am inclined to use it / be mindful of its necessity. The rhetoric of a punitive God is likewise seldom invoked. In this secular context, one hears more about themes like justice and responsibility, but I often feel that they are untethered from how we talk about God. Some of the Indigenous students I have taught would be wary of how some talk of providence and "God will provide" leads to passivity: there are occasional references to (as in Kiribati/Tuvalu) how talk of God's care and provide[nce] can take the form of a particular type of climate change denial.

ER: Clive, you are right, but in our essay, we suggested that even where God cares or provides for, human responsibility has a role to play. In our essay, we mentioned the Rwandan saying *"Uragiriwe n'Imana ashyiraho n'uwe mushumba,"* which means: "Even the one whose cattle is cared by God has to hire someone to look after the cows." This shows that in Rwandan mentality, passivity and laziness have no place in believing in God's providence. Believing in God's providence sometimes has been used as a way of escaping from responsibility, for example, by not taking required actions to curb environmental degradation. But in fact, God's providence urges us to protect God's creation.

NHC: Clive, I agree that a strong belief in God's providence can lead to a fatal passivity. In my context, the providence of God is constantly affirmed in the liturgy (Anglican) but is not really referenced in everyday life. Until the last few decades, to do so would put you into conflict with the major dictates of evolutionary biology. Now, that is less the case, especially if one is affirming the subtle actions of God in and through creation. It is hard to hold both the need to act and the importance of human action in tension with a strong understanding of God's immanent presence. The God of science was an overwhelmingly deistic God in the twentieth century, apart from holdouts of process theology. In process thinking, God's lure could often seem like a veneer that lay on top of a mechanistic world and could never be tested, much like the ether. This is beginning to change, but slowly.

EMC: By contrast, I need to admit that God's providence was an important undercurrent in the Dutch Reformed Church in which I grew up. Many of the hymns that we sang in full voice were about God's care. Alas, this had to with God's protection within a family lineage over many generations (in a colonial context!) and with a cultured nationalism where God's people ("*volk*", i.e., the Afrikaner *volk*) were the primary recipients of such providential care.

GvdB: And perhaps that is exactly why a volume like the present one is so welcome. Hearing all these different voices from so many places across the globe reflecting on the problems and prospects of providence-talk will hopefully forever prevent us from closing in on our own "tribe" when confessing providential God's care.

EMC and ULV: In a similar vein, would you agree that there is a tendency to reduce providence to a discussion of the intractable theodicy problem? And that some then focus on natural suffering while others focus on social sources of suffering? Does this distinction still hold in the "Anthropocene"? But one cannot avoid the theodicy problem either, don't you think?

ER: In my view, the distinction between natural and social sources of evil allows for condemning negative human behavior while reinforcing positive behavior amid the "Anthropocene." Even though the role of humans remains contested, its recognition allows for further reflection.

GvdB: Yes, I agree. The notion of providence inevitably evokes the theodicy problem, since there are so many situations in which God's expected and prayed-for care fails to show up—which immediately prompts the classical "why?" question. Yet I have found it very helpful and instructive that most authors in this volume shelve the theodicy-issue a bit, in order (as Eraste suggests) to first of all focus on patterns of *human* behavior with regard to the evils of the "Anthropocene," including patterns of human religious behavior. In a sense, it is ironic that precisely belief in God's providence, which in so many biblical stories empowers people and make them resilient in the face of natural and social evils, has now become a source of resignation and acquiescence for so many people, as a couple of essays in this volume point out. So that leads to the question how we can distinguish between adequate and inadequate, or spiritually healthy and unhealthy, uses of the appeal to God's abiding care for the world.

EMP: I don't think the theodicy problem can be avoided; indeed, I don't think it *should* be avoided, which is perhaps why—though I can acknowledge the concern here—I initially bristle a bit at the suggestion that providence would be "reduce[d]" by linkage to this intractable problem. This is because I (a) understand the intractability of speaking meaningfully about God's loving care in the context of suffering to be rooted in a protest against that suffering, and (b) see that protest as incarnating the most valuable ethicospiritual insight: God does not want human beings to suffer, as Edward Schillebeeckx among others so passionately insists. Perhaps this is a circuitous way of affirming an enduring link to the God who loves and cares (i.e., through this mode of negative contrast, as Schillebeeckx calls it), and perhaps one that on its own is too fragile, too thin, to suffice as an understanding of providence. But maybe [it] can serve as an opening for such an understanding? I am less ready than Gijsbert to describe a focus on human behavior as putting theodicy to one side; to my mind, anthropodicy dissolves into the theodicy problem. However, I agree that his concluding question is absolutely the right one. As for the distinction between natural suffering and that which has social sources, I have become increasingly interested in seeing what comes to light when we consider the possibility that this distinction has *never* really held water. Such an integrated "eco-social" perspective is, I think, one of the real gifts of ecotheology.

CP: The two are indeed closely linked and almost necessarily so in the "Anthropocene"—but I do think they need to be kept discrete as well. They are not the same, but in the "Anthropocene," the problem of providence—coupled with humanity's power, the great acceleration, extreme weather events, et cetera—flow into theodicy because of increased vulnerabilities, and the enhanced innocence of those peoples' wise carbon emissions are miniscule. It is no longer simply a discussion between a natural and moral suffering humankind. The extreme natural events often now have an all-too-human degree of attribution, and that "moral"/"ethical" dimension raises questions of causality, responsibility, and justice.

ULV: I don't think the issue of theodicy is shelved, depending on which angle one approaches the issue from. I believe it is woven into the conversation of providence. However, for many years we've been talking about theodicy, but theodicy driven by the Western philosophical tradition that sometimes draws a sharp distinction between God's work of salvation (redemption narrative) and God's work of creation (creation narrative). We all know that this is the influence of the Reformation tradition, which unfortunately subordinates the creation story to the redemption story. What we need is a more holistic take of providence that does not sideline the importance of the suffering of creation and at the same time does not eliminate the importance of salvation. This is where Indigenous holistic philosophies come in to ground a new theological narrative that is more biblical yet also speaks to the suffering of communities, in particular Indigenous communities.

NHC: The more integrated we understand the natural world to be, the greater the theodicy problem in many ways. The more we push back the threshold of disease and distress into the early history of life on earth, the more we have to grapple with why there seems to be an excess of evil at all periods. The more we affirm providence, the greater the problem of evil. I have argued, though, that recognizing the love and glory that are evident in the natural world allows us to accept the suffering at the same time, in faith, as yet unexplained misery. And to take comfort that God, as revealed in Christ, is on our side and on the side of all life.

EMC: Yes, Nicola, I warm to every word you say. Let me nevertheless put this bluntly: I am deeply suspicious of the considerable interest in natural evil in some scholarly circles in the Global North. This may well sound like an apology for social evil. Something like: the Devil made me do it! Or, if not, my genes made me do it! By contrast, while natural sources are suffering are accepted as part of life in the Global South, the focus is very much on social evil.

ASA: As mentioned above, the theodicy problem seems to be one of the primary reasons why both critics and curious religious worshippers are hesitant to discuss the doctrine of providence. At the risk of exaggeration, I may state that addressing the theodicy problem is the core motif of the doctrine of divine providence. I think the dividing line between providence and responses to the problems of suffering (including ecological woes) is very slim. This is because divine providence encompasses God's love and care, governance, and preservation/sustenance of the creation. In that case, one cannot discuss the doctrine of divine providence without reference to a theodicy.

EMC: Aku, you may be right that discussions of God's providence are often prompted by the problem of suffering, but I would warn against reducing providence to theodicy, basically in order to avoid an intellectualizing of trust in God's care.

MS: I think that, yes, there is a tendency to reduce providence to theodicy, although mostly in theological approaches of a speculative-philosophical character. On the other hand, in the colonial imposition of Christianity in South America, there has been an "abuse of the cross," and in this sense I agree with Ernst about the dangers of emphasizing the vein of natural suffering. However, the various political and theological liberation movements in our region have reached a significant critical mass in recent decades. Indeed, the "naturalization" of suffering has always been a problematic issue in subjugated populations, but I observe in many local Christian communities a clear distinction between sufferings inherent in natural processes and those originating in social injustice. For this reason, I believe that the theme of God's providence is often set aside in order to avoid an uncritical acceptance of social suffering. In turn, "natural" suffering is seen as part of life. Theodicy thus has a very narrow field, because suffering due to injustice questions human sociocultural and political structures, not God. On the other hand, ecology in our region was until recently considered a "First World" intellectual luxury. This has changed only in the last few decades, and in the field of theology, Indigenous communities have contributed much to this change because of their holistic worldviews, an aspect that Upolu observes well. Then, as to the distinction between natural and social suffering in the current context, I would say that the boundaries appear clear in some respects and blurry in others.

EMC and ULV: What were the crucial insights that emerged for you as you encountered the essays of the other authors and worked on your own essay?

ASA: One of the crucial elements that emerges while reading other contributions is to find out that the doctrine of divine providence is less frequently discussed in other denominations. In the African context, all of life is saturated by the awareness of either God's providence or of fate. Further, it was also encouraging to me to see that other colleagues see divine providence extending to animate and inanimate objects. Clive's essay on "The Challenges of 'Angry Weather'" resonates well with my context. Although everyone seems to have a problem with unfavorable weather in the current secularized Europe, I doubt that Europeans see such problems with weather as having anything to do with the divine.

GvdB: Yes, I think you are right here, Aku—we very much live in a disenchanted world. Yet as a Reformed theologian from Western Europe, I am inclined to see notions of divine care and providence as important theological themes (rather than pieces of obsolete dogmatic lore). And it actually struck me that some colleagues from other denominations and other parts of the world are far less at ease with them. I suppose this is not just a matter of having had a different sort of theological training but also of these colleagues being led by a hermeneutics of suspicion that was fueled by the massive abuses that have been made and are being made of these notions in so many social and relational contexts. Having been aware of the specific historical twists and turns of the doctrine in the West (e.g., its being co-shaped by Stoic notions of fatalism and determinism), at least since the time of my doctoral work, reading the essays of the other contributors to this volume—especially the non-Western ones—impressed me with the extent to which our hybrid (i.e., partly biblical, partly Greek-philosophical) Western notion of providence has spread across the world as a "successful" by-product of Western colonialism—with all shadow sides that came with it. But perhaps in non-Western cultures it came to tie in with deep-seated intuitions of fate and necessity that already abounded in their pre-Christian pasts.

CP: I find reading other essays a challenge in the light of seeking to think in terms of providence when the issues may be presented through cultural worldviews, other disciplines, events, and programs, and at times a relative absence of talk re providence. I think it is a tension to wrestle with because I think that the Christian faith does have a "theological agenda" (in terms of issues/beliefs that necessarily arise out of faith, independent of the particularity of cultures), and I have increasingly thought that the Christian faith itself is not unlike a distinctive, other type of "culture" that can be sidelined.

ER: On a personal note, I may say that reading the other essays has opened my mind. Your ideas have shaped mine so that I could taste the complementarity of being human. It helps me to listen to stories about God's work from elsewhere in the world and to find that then also in my own life and culture.

NHC: A word on such encounters beyond this volume: Aotearoa / New Zealand is a very secular country, and wherever secularity is high, there is a reluctance on the whole in the church to talk about providence. On the other hand, we are in close conversation with Mātauranga Māori (Māori ways of knowing), and this Indigenous outlook assumes a holistic, spirit-imbued vision of nature, including an understanding of *whakapapa* (genealogy) of all things in some spiritual source. This Indigenous thinking and the Christianity associated with it push us towards a deeper embrace of what we call providence.

GvdB: It is interesting to see how secular societies that are fed up with traditional Christian notions can start tapping into other (partly similar) resources to fill up the gap, as it were. Thanks for pointing out how this works at your place, Nicola. We see similar patterns here in Western Europe. Of course, we don't have an Indigenous minority population, but rapprochements with Eastern religions seem to play a similar role; for example, *karma* is by now a very popular word/concept among young people here.

EMC and ULV: "The making room" in the title of this volume suggests that God's providence is not necessarily an aim in itself but that the story needs to continue. In that sense, God's providence is a response to a story that has gone wrong. What do you make of that?

NHC: I think we are at a time in history where our theology, our biology, and our physics are going through paradigm changes and also becoming more entangled narratives. Even without the "Anthropocene," then, this will open up new chapters in the narrative of reflection on theodicy and providence, hence making room for the story to continue.

EMP: For me this is a provocative prompt—in a good way! It spurs me to look back at Gijsbert's comments in response to question 2, when he notes that the "cradle" of belief in providence has been situations of "utter despair"—where the story stalls and sputters, we might say. This only redoubles the original question, though, for is providence then just one contingent response to an experiential and/or narrative breakdown? Or is it the ongoing name for this aspect (viz. enduring trust) of the relationship between God and the world, or God and God's people? Could Christians tell their story without it? One of the assets for me in engaging other contributors' work in this volume is being able to sort of play with these big and important (... and abstract) questions while not losing sight of the particular content of this doctrine, disputed as it may be, and how wrestling with it can bear fruit in our respective contexts and our interpretations of this planetary condition.

GvdB: I agree with you, Elizabeth! It is indeed a sometimes frustrating but also a fascinating nexus of questions that is opened up here. And the way we deal with them certainly matters. (Personally, I'd suggest there is back-and-forth between deeply existential experiences of providence in times of crisis and more "stable" and enduring notions of it, which should not be cut loose from the doctrine's existential origins.)

ULV: I also like the idea of a "response" to the "Anthropocene" story, or any other story gone wrong.

MS: Yes, Upolu, it seems to me that we don't know of any example of a story "going right." The mythic narrative of the fall in Genesis is the theological correlate of an already fallen reality. The only experience we have is that of an "imperfect" story in which suffering has always played a dominant role. It is interesting that in this theological mythical narrative, the catastrophe begins with the offer of knowledge to become like God, an offer that does not come from God. Anyway, as human beings in general and as theologians in particular, we try to give answers to what we do not know. I do not know if we are the ones who "make room"; the history of this planet and the evolution of life on it are too big to think that we hold all the coordinates. Yet we should make *responsible* efforts, animated by faith—which, for me, is the ground of providence.

EMC: Thanks, Marisa, for this emphasis on who is making room. It would be arrogant to suggest that we (as humans) have to make room for God's work. It would be even worse to think that we (as theologians) would need to make room for the story that we are telling to continue. In this sense, "making room" is indeed God's work—as seen retrospectively through the eyes of faith.

CP: Making room to me suggests that providence may well have been squeezed out in responses to the "Anthropocene"—probably out of a concern for theodicy and the need to address questions of justice. So often, providence features in Christian thinking like wallpaper in a room: it is barely noticed, seemingly squeezed out … but is the kind of doctrine that permeates the whole of theology: so it is part of the picture, part of the response, and maybe one that needs a little bit more assistance to be noticed.

ASA: Depending on one's position, a supralapsarian, in contrast to an infralapsarian, might not see divine providence as "a response to a story that has gone wrong." Many contemporary scholars, especially open theists, may rightly see the *need for the story to continue* because of their view that creation is an open project with open routes. These open routes allow humans to influence creation positively and negatively. In that case, divine providence might be rightly conceived as "a response to a story that has gone wrong." Instead of seeing divine providence from the latter perspective, I would want to see the continuation of the story from the perspective that God changes strategies in relation to human input. God's love and care for creation is demonstrated constantly in response to preserving and sustaining the creation despite destructive human impact.

ER: This volume has indeed opened a new page so that everyone is invited to reflect on the story of God's work. This will shape attitudes to the natural environment for better or for worse.

EMC and ULV: In some theological traditions, three or four aspects of God's providence are distinguished, namely ongoing creation, conservation, governance, and concursus (the interplay between divine, human, and other forms of agency). Do you find this distinction helpful or not? Where was the focus of your essay?

ER: This distinction is indeed very important because it allows for different ways of experiencing God's work. In our context, it is especially God's work of conservation that matters.

GU: Indeed, the interconnections between creation, environmental protection, governance, and the role of people require sustained further reflection. This is not the task of individuals but requires collaboration and partnerships between churches, governments, NGOs [nongovernmental organizations], and schools.

CP: Yes, Eraste, the typology is helpful, but it also makes me despair. It seems to me that in the "Anthropocene," humankind—and particular powers—is/are exercising governance at the expense of *concursus* and God's *gubernatio*. I think the Christian narrative can provide a valuable counter-narrative/foil—a reminder of a calling that goes beyond references to stewardship, relationship, et cetera. These things are there and important, but the typology places the divine more overtly into the situation. I find myself wondering how the threefold typology accommodates Christology—rather than often being an extension of the doctrine of God, creation, and a continuing creation.

GvdB: I think such distinctions are certainly helpful, since otherwise our thinking on providence could easily remain quite vague and superficial (perhaps even dangerously so). Yet it seems important that such distinctions and divisions don't narrow down the many shades and shapes of God's providence as narrated in the Bible to only a couple of them. In fact, the focus of my essay was to start from this classical trio (which, by the way, nobody knows exactly where it comes from or who invented it) but then to open it up towards a wider view which also includes interrupting, liberating, and transformative patterns of God's loving care for creation.

ASA: I also agree with Gijsbert that these distinctions are helpful. In other to avoid being repetitive, I did not discuss these distinctions in detail, but they form the basis of my discussion: the distinction between conservation, governance, and *concursus* was assumed in my paper.

EMC: Fair enough, Gijsbert. I fully agree, also, with David Fergusson's "polyphonic" approach. In my essay I deliberately focused only on the theme of conservation (understood as common grace) or in secular terms the theme of sustainability.

EMP: By contrast, as a theologian trained primarily in the Catholic tradition, these distinctions were not especially familiar (at least as a set), but hearing the discussion around them has been interesting. I can see the utility, though have the sense that they are helpful only so far as they go … and, in the context of this volume's key question, I'm not entirely sure how far that really is. For me, the point Ernst names in conjunction with question 1 continues to be the most important: namely, that God's work of salvation not be separated from God's work of creation. I am also curious whether the distinction between conservation and governance is one that replicates a distinction (verging on a separation?) between "natural" and "social" spheres and, if so, how challenges to the latter (natural vs. social) might reflect back on the wisdom of maintaining this particular theological distinction.

ULV: As a Methodist, we don't have these distinctions, so they are unfamiliar. However, from an ecumenical perspective, these traditions are important for me. I would caution that in many of these Christian traditions, they somehow compartmentalize the narrative, especially when it comes into the wrong hands, in particular those who do not understand these traditions. That is why I think I would be cautious to write on these.

EMC and ULV: How, then, would you respond directly to the core question of this volume: "How could the suffering of God's creatures in the 'Anthropocene' be reconciled with trust in God's loving care?" Does God's loving care still make sense?

GvdB: When confronted with such direct yes-or-no questions, as a theologian I am usually a bit hesitant, since one easily ignores much-needed nuances when answering them in any straightforward way. But limiting myself to the second question here: yes, I certainly do believe that trust in God's loving care still makes sense, insofar as it is based on the promises of the gospel. As such, it does not lead us into the temptations of passivity and complacency but encourages us to fight for a world which at the end of the day continues to be God's world. Isn't that what the doctrine of divine providence (and especially, perhaps, its notion of conservation) expresses at its deepest level: that our threatened world continues to be a world God is working in and fighting for, so that we have ample reason not to give up on her either?

EMP: I would want to distinguish "making sense" in a theoretical or rational way from a kind of fidelity to a conviction that is more experiential and praxis-based, where the "sense" or meaning at hand is of a kind that resists rational closure—and I suppose narrative closure, as well—but does not, on that basis, simply reduce down to nonsense. If I'm reading Gijsbert correctly, this view seems to be consistent with his powerful closing line just above.

EMC: Well said, Gijsbert and Elizabeth! One may say that God is like a parent fighting for the life of a beloved child.

NHC: Yes, indeed, I don't think faith will continue if we don't reassert providence and a belief in God's loving care. But this love was never promised without suffering—for some reason. The doctrine of providence will flourish if science and faith are pondered together, and if the doctrine of the Spirit and incarnation is also upheld and reflected upon.

ASA: In my discussion on the problem of suffering, I have come to realize that although we could provide viable responses to the problem of ecological woes in the "Anthropocene," such responses might make little sense to some, especially those who do not believe in the existence of a loving God as conceived in Christian theism. However, as Gijsbert stated, I think it is tricky to provide a black-and-white answer as to whether God's loving care still makes sense today. Our beliefs will definitely determine our responses.

ER: This question raises debates that cannot be resolved. Does God have no power to stop the suffering of creatures? Is suffering the result of God's punishment or God's pedagogy? Yet God's love and care still make sense. The collection of essays in this volume attests to that.

MS: In my essay, I have the premise that the theme of providence belongs to the realm of faith and not to that of rational speculation. The trust in God's loving care in the face of suffering rarely "makes sense." So any responsible action in this world and for the sake of others is always an act of faith, a going out of ourselves in response to a promise of a future that we expect but that we do not know.

EMC: Ah, the apophatic!

EMC and ULV: This series seeks to optimize diversity in ecotheology. Amid such diversity, what "current paths" are you able to identify in the field? And what "emerging horizons"? Where is the debate going? What work is still needed?

EMP: Just a few scattered comments on this collection of weighty questions. First, it does seem that ecotheology has proven to be an area in which real strides can be made toward displacing the hegemony of Euro-American perspectives. Even as I say this, I worry it's an overly hopeful reading. Yet there do seem to be features of ecotheology (e.g., its global reach; the nature of the challenges and boundary-pushing questions it poses) that facilitate some reconfigurations of power and voice concerning the issues that matter for Christians—and those that should matter more. My sense is that the theme of ecojustice has become a path that anyone writing in this field must tread in some way in order to make a meaningful contribution; this is a good thing. Questions surrounding posthumanism and transhumanism strike me as an emerging horizon; typically, it has been a fraught one for telling the Christian story, especially when such questions veer into species extinctionism and arguably misanthropic evaluations thereof. To be sure, certain inquiries in this vein simply re-instantiate a myopic tendency on the part of elites in the Global North. Yet are these kinds of issues necessarily at so many removes from the concerns grouped under the umbrella of ecojustice? And/or of the concerns of non-Western and Indigenous peoples? Must they be addressed given what is at stake in the "Anthropocene"? I'm genuinely not sure and see this as an arena for further work in the field.

ULV: There are definitely emerging horizons. One obvious one is Indigenous knowledge and philosophies. One of the pushbacks from Indigenous theologians in terms of theologizing is the compartmentalization of theological concepts and categories that leads to splitting one from the other. I believe that there is much work needed in such Indigenous theologizing, in particular in its commitment to address the "Anthropocene" from a different perspective. God's work of creation definitely needs to be deeply connected to Indigenous, earth-oriented philosophies and knowledge.

NHC: Upolu, I would agree that ecotheology (with its strong links to feminism) is important in widening the conversation, as is the conversation with Indigenous theology. I would also argue that the conversation with the paradigm changes in evolutionary theory is absolutely needed to write a coherent narrative of faith and science. In this field, I would say that Celia Deane-Drummond excels in all of these directions.

CP: I have been working in this field since the late 1980s. I no longer speak of ecotheology. I talk about theology and climate change / theology and the "Anthropocene" / theology and Earth Systems science. I am not a scientist myself, but I do think the discussion needs to be interdisciplinary and engage with leading Earth System scientists (who often see the COP [United Nations Framework Convention on Climate Change Conferences of the Parties] conferences as conservative) and how our thinking engages with geopolitics.

EMC: That's interesting, Clive. I am also wondering whether the emphasis on "*oikos*" in ecotheology still holds given the notion of planetary boundaries in Earth System science. I am not drawn to "planetary theology" either, perhaps because it is devoid of symbolic power.

EMC and ULV: The title of this volume suggests the need for the story of God's work to continue. A discussion of God's providence is "making room" for God's work of salvation (Volume 5), the formation of the church (Volume 6), and consummation (Volume 7). How do you see the story unfolding given the contributions to this volume?

NHC: An important associated question is what does salvation mean?

ER: The wish to make room has now been expressed and the continuation of the story of God's work remains a *sine qua non* for the church. Our task is to sow good seeds and let God do God's work.

GvdB: To be honest, even though I am deeply convinced of the narrative structure of all proper Christian God-talk, and I think this book series will recontextualize the story of God's work in powerful ways for the present age, I am not so sure whether this can be done in a couple of neatly divided segments. In any case, the present volume seems to show that the notion of God's providence, when geared to its manifold biblical uses, defies traditional attempts to "store" it in discourse on the first article only. Instead, God's providential care for humans and the world is also (or even mostly, as the late German theologian Christoph Schwöbel has somewhere suggested) realized through God's acts of salvation and consummation, God providing salvation for us by God's two "hands," Christ and the Spirit. In such ways, Christian notions of creation, providence, and salvation, though each having their own part to play, seem closely interconnected.

CP: At some stage, I think there needs to be a deeper probe into the nature of what it means to be human in the "Anthropocene" and how we are to understand the person and work of Christ with regards to climate change / loss of biodiversity, et cetera. I am haunted by a line I discovered in preparing my chapter: "Where were you when I went extinct?" It's a question which represented an extension of Matthew 25.

EMC: I guess it is then a matter of "Watch this space!"

Key

ASA:	Aku Stephen Antombikums	GU:	Gloriose Umuziranenge
CP:	Clive Pearson	GvdB:	Gijsbert van den Brink
EMC:	Ernst M. Conradie	MS:	Marisa Strizzi
EMP:	Elizabeth M. Pyne	NHC:	Nicola Hoggard Creegan
ER:	Eraste Rukera	ULV:	Upolu Lumā Vaai

Index

A
Abraham, 20, 56, 59, 62, 67-68, 79, 86, 222-223, 225, 228
Accra Confession, 75, 77
adaptation, 77, 101, 144, 197, 202, 205, 211, 236
African worldview, 46, 24
Alves, Rubem, 30
angry weather, 125-126, 128, 130-136, 138, 140-142, 144-145, 150, 245
Anthropocene, 1-3, 5, 10, 12-16, 22, 26-30, 35-38, 44, 48, 51, 55-56, 58, 73, 75, 79, 82-84, 86, 89-92, 94-97, 99-101, 103-104, 107-108, 111-113, 116, 122-123, 125-129, 135, 137-145, 149-150, 152, 154, 156-175, 181-182, 184-187, 190, 193, 196, 199, 201-202, 207-208, 210, 215, 217-218, 220, 226, 230, 235-240, 242-243, 247-253
anthropocentrism, 17, 162, 165, 170
anthropodicy, 155, 243
anthropogenic climate change, 12, 18, 23, 144, 232
Aotearoa, 89-90, 92, 94, 96-100, 102, 104, 106, 108, 246
apartheid, 16, 20, 60, 68-70, 221, 240
Aquinas, 39, 151, 224
Argentina, 181, 183-184, 197

B
Barmen Declaration, 67
Barth, Karl, 30, 86
Bavinck, Herman, 86
Bergmann, Sigurd, 145
Berkhof, Hendrikus, 30, 232
Berkouwer, Gerrit C., 30, 86
Bible, 37, 40, 51, 59, 80-81, 83, 89, 135, 168, 185, 204, 216, 220-225, 229, 249
biodiversity, 13, 21, 23, 145, 161, 208, 217-218, 220, 238, 253
biosphere, 173, 216, 218, 229
Bonhoeffer, Dietrich, 30, 86
Brundtland report, 21

C
Capitalocene, 10, 16
carrying capacity, 21, 74
catastrophe, 22-23, 25, 132, 160, 199, 203, 220, 237, 248
ceramic water filters, 206
Chakrabarty, Dipesh, 30
Clark, Nigel, 30
climate change, 5, 12, 18, 20, 23, 30, 36, 41, 46-51, 76-77, 84, 101, 108, 126-135, 138-139, 141-145, 156, 159-160, 162-163, 169, 171, 173, 183-184, 197, 199-211, 217-221, 230-232, 241, 252-253
climate denialism, 218
climate despair, 232
climate indifferentism, 218
colonialism, 10, 17, 22, 131, 138, 142, 168, 172-173, 246
colonization, 27, 238
common grace, 4, 13, 20, 55-66, 68, 70-72, 74, 76, 78-80, 82, 84-86, 226-227, 250
concursus, 3, 15-16, 18-19, 21, 23, 30, 36, 58, 73, 84, 86, 129, 131, 141, 143-144, 153, 166, 207-208, 210, 226, 231-232, 249
Conferences of the Parties (COPs), 202
Conradie, Ernst M., 211
conservation, 15-16, 20-21, 27, 36, 58, 66, 72-73, 84-86, 121, 208, 224, 229, 232, 249-250
consumerism, 6, 77-78
consummation, 13, 29-30, 73, 84, 144, 224-225, 229, 253
Cornwall Alliance, 220, 231
coronavirus disease 2019 (COVID-19), 11, 183
corruption, 9, 78
cosmos, 38-39, 43, 51, 64, 79-82, 97, 182
courage, 7-8, 18, 82, 86, 223, 232
covenant, 7, 14, 17, 49, 56-57, 59, 63, 65, 78, 84, 86, 127, 230
cows, 215-216, 218, 220, 222, 224, 226, 228, 230, 232, 241
creatio continua, 3, 15, 153, 163
creatio ex nihilo, 38, 4
cross, 6, 10-11, 30, 62, 108, 112-113, 115-116, 120, 144-145, 156, 166, 188-190, 192, 196, 211, 228, 232, 237, 245

D
damnation, 97
das Nichtige, 67, 8
Davis, Heather, 145
Dawson, Ashley, 145
decolonization, 24

deep incarnation, 28
democracy, 24, 81, 101
denaturalizing, 172–173
determinism, 5, 8, 15, 41, 73, 166, 224, 246
development, 14, 21, 23, 26, 30, 58, 69–70, 72–75, 77–79, 103, 133, 139, 164, 181, 183, 206, 215, 220, 224, 236, 239
displacement, 17, 118, 171
disruption, 7, 26, 83, 86, 90, 96, 100–101, 104, 107, 118, 156, 227
divine determinism, 5, 73, 224
divine election, 20, 85
divine foreknowledge, 8
divine justice, 11, 166
divine pedagogy, 208
divine sovereignty, 50
divinization, 28, 63

E
Earth System, 2–3, 22, 74, 86, 96, 104, 126–127, 134–135, 137–141, 145, 159–160, 164, 171–174, 238–239, 252
ecofeminist theology, 18
ecojustice, 23, 58, 75–78, 86, 168, 252
ecomodernism, 145, 167, 175
Ecomodernist Manifesto, 139, 145, 165, 170
economic growth, 10, 23, 27, 74–75, 77, 86, 211
Ecumenical Patriarchate, 202
energy, 76, 79, 82, 99, 144, 205
Enlightenment, 23, 27–28, 37, 102–103, 165, 168
eschatological consummation, 30, 224
eschatology, 14, 25, 44–45, 47, 61, 72, 131, 218
eschaton, 51
Eucharist, 98
evil, 4, 6, 9, 11–12, 14, 22, 25, 27, 29, 35–38, 44–45, 48–49, 51, 56–57, 60–62, 65, 68–73, 79, 83, 85, 90–91, 93–96, 102, 108, 126, 130, 143, 150–151, 156, 182–184, 186, 189, 191, 196–197, 224, 230–231, 236, 238–239, 242–244
evil spirits, 35–36
evolution, 19, 22, 27–28, 66, 89–90, 92–94, 96, 98, 100, 102–108, 166, 181–182, 248
extinction of species, 208
extreme cities, 140–141, 145

F
farming industry, 217
fatalism, 8, 39, 246
fecundity of life, 12, 18
fideism, 96
food security, 16, 205
fragments, 118, 229

G
Gesetz der Umlenkung, 6
Global North, 58, 77, 79, 86, 138, 150, 217, 220, 244, 252
Global South, 17, 22, 24, 37, 58, 77, 79, 86, 138, 168, 205, 244
gnosticism, 26
God as Creator, 11
God as Savior, 12
God's care, 5–6, 8–11, 13–14, 16–18, 20–22, 26, 63, 84–85, 153, 167, 207, 216–218, 221–222, 225, 230–232, 237, 239–242, 244
God's covenant, 7, 49, 56, 59, 78, 84, 230
God's economy, 85
God's finger, 25
God's governance, 10, 15–16, 19, 24–25, 27, 29–30, 36, 58, 63, 86, 170
God's hiddenness, 154, 187, 189
God's identity and character, 2
God's mission, 13, 56
God's omniscience, 4
God's patience, 13
God's sovereignty, 10, 38, 78, 131
grace, 4, 6, 11, 13, 20, 50, 55–68, 70–72, 74, 76, 78–80, 82, 84–86, 94, 145, 151, 153, 155–156, 191, 195, 197, 217, 226–227, 250
great acceleration, 144, 243

H
Hamilton, Clive, 108
Haught, John, 108
healing, 17, 41, 100, 156, 168, 236
Heidelberg Catechism, 7, 224, 232
heteronormativity, 24, 7
Hiroshima, 22, 38, 196
HIV, 11, 17
Holocene, 3, 27–29, 37, 56, 58, 73–75, 79, 82–83, 86, 90–91, 94, 97, 108, 126–127, 135, 137, 143–145, 159, 207
Holocene stability, 58, 73–75, 79, 82, 86, 91, 94, 108
Holy Communion, 80
Holy Spirit, 78, 82, 115, 192–193, 228, 231–232
homo deus, 28, 86
homo excelsior, 28
hope, 6–7, 21, 25–26, 28–30, 44, 51, 57–58, 60, 70, 84–85, 103–104, 120, 142, 145, 153–154, 156, 159, 192, 201, 208, 211, 232, 237
hothouse, 134–136, 145
household of God, 76
human agency, 18, 51, 91, 140, 144, 164–166, 175, 182, 194, 222, 226
human dominion, 193

human freedom, 40, 42, 46, 49-50, 73, 152, 154, 227
human immunodeficiency virus, 11
human rights, 24, 184, 197
human sin, 4, 9, 14, 20, 22, 29, 57, 59, 66, 79, 190, 224

I
idolatry, 77, 98, 104, 191
Indigenous communities, 5, 18-19, 239-241, 244-245
Indigenous theology, 17-18, 207, 252
Indigenous wisdom, 4, 17
industrialized capitalism, 10, 29
Intergovernmental Panel on Climate Change (IPCC), 84, 129, 159, 200
interreligious dialogue, 225
intersectionality, 136, 138

J
Joachim of Fiore, 71
joy, 65
judgment, 57, 83, 131, 170, 185-186, 227, 231
justice, 11, 22-23, 42, 45, 59, 70, 72, 75-79, 83-85, 104, 132, 137-139, 145, 166, 172, 186, 188, 197, 201-202, 206, 208, 210-211, 238, 241, 243, 248
justification, 12, 71, 116, 157, 170, 187, 197

K
kairos, 26, 70, 200-201, 211, 238
karma, 8, 225, 247
kenosis, 28
kingdom of God, 25-26, 29, 72, 228
Kiribati, 76, 241

L
laity, 5
liberation, 7, 13, 15, 23-24, 26, 68, 71-73, 83, 98, 107, 137, 150, 164, 172, 216, 224, 226, 228, 245
liberation theology, 15, 23, 68, 72, 98, 137, 150
limits to growth, 21, 73
Lisbon earthquake, 22, 38
love, 5, 7-8, 14, 29, 38, 50-51, 63-65, 82, 84, 90, 93, 98-99, 103, 117-118, 135, 145, 151-152, 172, 186, 188, 195-196, 202, 235, 238, 240-241, 243-244, 248, 251
luck, 8, 10, 145

M
Manicheism, 83
materialism, 24, 90-92, 95-96, 102-105, 238
memory, 56, 112-113, 119
Messiah, 8, 56

metaphysical evil, 11
meteorology, 136
miracles, 37, 63
mitigation, 76-77, 144, 202, 211

N
narrative, 2-3, 15, 68-70, 80, 91, 95, 102, 107, 121-122, 127, 130, 132, 134, 141, 149, 155-156, 158, 160, 162-164, 166, 170-172, 174, 182, 193, 204, 207-208, 222, 225, 227-229, 235-236, 244, 247-250, 252-253
natural evil, 4, 38, 73, 244
natural selection, 12, 96-97, 99, 106, 172
natural suffering, 11, 57, 242-243, 245
nature conservation, 20-21, 73, 121
Nazi Germany, 16, 2
neo-Calvinism, 59-60
new economics, 24
Noah, 17, 55-56, 69, 86, 127, 230

O
ocean acidification, 13, 23, 38, 161
omnipotence, 60, 63, 84, 150, 154
open theism, 36-38, 40-41, 50-51
optimism, 24, 38, 84
orders of creation, 58, 60, 66-71
ordinances, 16, 20, 65-67, 70-71, 216
Ordungstheologie, 66-67
Orthodox theology, 14, 21, 73

P
pandemics, 11, 17
Pantheism, 13, 19, 152
Paraclete, 30
parousia, 47-48, 51
Pasifika, 15, 76, 82
pastoral care, 17-18, 142
patriarchy, 18, 70, 138
peace, 59, 65, 72, 75, 78-79, 104, 156, 190, 202
Pentecostalism, 38, 41-42, 46, 49, 51
planetary boundaries, 21, 74, 145, 159, 252
Plato, 141
play, 2, 9, 23, 30, 36, 39, 57, 59, 80, 82, 99, 127, 129, 133, 143, 155, 166, 184, 241, 247, 253
posthumanism, 252
predation, 22
predestination, 50-51, 221
predicament, 10, 42, 44, 46, 127, 137, 139, 142, 145, 157, 173, 175, 218, 223, 236
preservation, 20, 61, 67, 71, 73, 85, 144, 211, 224, 226, 229, 244
progress, 14, 23-24, 72, 75, 90, 105, 131, 165, 170

pronoia, 222
proton, 30
punishment, 11, 14, 43–44, 46, 48, 135, 162, 190, 208, 239–240, 251

Q
quasi-soteriology, 69

R
re-creation, 62–64
resignation, 7, 70, 162, 197, 243
resilience, 18, 92, 108, 120, 144, 159, 200, 205–206, 211, 229
resistance, 7, 23, 90, 104, 111–114, 116, 118–120, 123, 131, 222
responsible stewardship, 21
resurrection, 7, 30, 144, 156, 196, 230
Rio Earth Summit, 73
rupture, 2–3, 94, 126, 138–140, 167, 169, 174, 207, 217, 237–238
Rwanda, 130, 199–203, 205–208, 210–211, 241

S
sacred, 41–43, 45–46, 49, 51, 92, 119, 216
salvation, 4, 6–8, 13, 16, 20, 23, 26, 30, 40–42, 47, 49, 51, 56–57, 60, 63, 68–69, 72–73, 79–80, 84–85, 92, 99, 103–104, 150–156, 163, 224–225, 229, 236–237, 244, 250, 253
sanctification, 49, 71, 103, 197, 224
Second World War, 221, 227
secular, 8, 15–16, 21, 23–25, 28, 37, 58, 72–73, 97, 127, 131–132, 153, 155, 162, 170, 232, 241, 246, 250
secularism, 26
serenity, 7, 8
slavery, 6, 10, 24, 119, 197
social evil, 4, 22, 73, 243–244
social suffering, 245
South Africa, 1, 16, 20, 25, 35, 59–60, 66, 76, 203, 215
species extinction, 21, 218
Spirit of God, 62, 95
stirring the soup, 82–83, 86
Stoicism, 28, 131
structural violence, 11, 22–23, 57, 79, 173
survival, 82, 86, 116, 121, 126, 165, 200
sustainability, 21, 23, 55–56, 58, 60, 62, 64, 66, 68, 70, 72–78, 80, 82, 84, 86, 94, 211, 226, 250
sustainable development, 21, 23, 58, 73, 75, 77–78

sustainable development goals (SDGs), 75
sustainable growth, 23, 30, 73
sustainable livelihoods, 17, 21, 73
sustainable society, 21, 73, 75
sustenance, 6, 15–16, 21, 85, 90, 113, 244
Synod of Barmen (1934), 67

T
technology transfer, 77, 202
teleology, 25, 28–29
theocentrism, 44–45
theodicy problem, 4, 12–14, 22, 45, 85, 125, 242–244
theology of glory, 9, 188
theology of history, 26, 30
theosis, 28, 86
thermodynamics, 19, 58
tipping points, 91, 104, 136, 142, 159
tornado, 129, 131, 135–136
Totius (Jacob Daniël du Toit), 68
tradition, 2–4, 13, 15, 20, 36–37, 40–41, 55, 58–59, 64, 84, 91–92, 97–100, 102, 107–108, 117, 123, 126, 144–145, 149–150, 154, 201–202, 215, 223–224, 229, 238–241, 243–244, 249–250
tragedy, 14, 57, 86, 94, 142, 151, 236
transformation, 24, 70–72, 78, 106–107, 211, 226, 230
Trinity, 15, 64, 111, 193, 221, 226
trust, 1–10, 12, 14, 16, 18, 20, 22, 24, 26, 28, 30, 36, 55, 59, 61, 67, 84, 112, 116, 118, 120, 151, 153, 155–158, 163, 175, 185–186, 195–196, 201, 217, 219–221, 223, 225, 230–232, 237, 239, 244, 247, 250–251
Tuvalu, 76, 241
tyranny, 9, 11, 61, 70

U
universe story, 18
utopia, 24, 71

V
vicarious suffering, 11

W
weather attribution science, 132
web of creation, 138, 216–217
wisdom, 4, 6, 8, 17, 19, 29, 90, 95, 99, 103, 105, 107–108, 112–113, 119–121, 144, 153, 188–190, 220, 227, 250
World Council of Churches (WCC), 21, 75, 200

www.ingramcontent.com/pod-product-compliance
Lightning Source LLC
Chambersburg PA
CBHW081145230426
43664CB00018B/2813